Contents

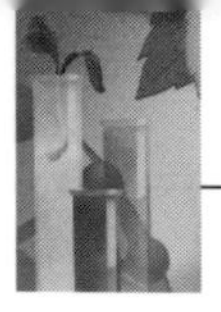

Introduction

About this guide

The aim of this guide is to help you prepare for the **A2 Unit 2: Biochemistry, Genetics and Evolutionary Trends** examination for CCEA® biology.

The guide has three sections:
- **Introduction** — this provides guidance on the CCEA® specification, and offers suggestions on improving your study/revision skills and examination technique.
- **Content Guidance** — this summarises the specification content of A2 Unit 2.
- **Questions and Answers** — this provides two exemplar papers for A2 Unit 2 for you to try. There are answers written by two candidates and examiner's comments on each candidate's performance and how it may have been improved.

The specification

You should have your own copy of the CCEA® biology specification. This is available from **www.ccea.org.uk**.

A2 biology

The A2 biology course is made up of three units. A2 Unit 1 and A2 Unit 2 (each worth 40% of the A2 marks) are assessed by examination papers. A2 Unit 3 (worth 20% of the A2 marks) is assessed internally on the basis of the investigational and practical work (coursework) that you will do in biology classes. The A2 units count for 50% of the overall A-level; the other 50% is accounted for by the AS units.

Links with AS units

A2 biology often builds on learning at AS. For example:
- to understand how genes function you need to understand the 'structure of nucleic acids'
- an understanding of gene technology builds on the tools and techniques described in 'DNA technology'
- evolution and speciation overlaps with the 'role of selection in maintaining the adaptiveness of populations of organisms in their environment'

Links with AS topics are given throughout the Content Guidance section.

Assessment objectives

Examinations in A2 biology test three assessment objectives (AOs). AO1 is about remembering the biological facts and concepts covered by the unit. AO2 is about being able to use the facts and concepts in new situations. AO3 is called How Biology Works. It emphasises that biology, as a science, develops through testing hypotheses.

In A2 papers there is a greater emphasis on showing your ability to *apply* your knowledge and understanding.

The A2 Unit 2 paper

While the A2 Unit 1 paper is available in both the spring and summer examinations, the A2 Unit 2 paper can only be taken in the summer. Therefore, it will probably be your last exam in A-level biology.

The A2 Unit 2 examination lasts 2 hours. The paper consists of about nine questions; the mark allocation per question is from 3 to 18 marks. There are two sections. In Section A all the questions are structured. In Section B there is a single question, which may be presented in several parts, and which should be answered in continuous prose.

Questions towards the start of the paper and the initial parts of questions assess 'knowledge and understanding' (AO1). There will also be questions that present information in new contexts and may test your skills in analysing and evaluating data (AO2).

At A2, AO3 may be assessed by questions that ask you to formulate a hypothesis, design an appropriate experiment, evaluate the methodology used in an experiment and analyse results to determine whether a hypothesis has been supported or disproved. You may also be asked to consider the ethical implications of the way in which research is carried out, in which case you are expected to make informed (not emotional) comment.

You are expected to use good English and accurate scientific terminology in all your answers. Preparing a glossary of terms used in each topic should aid this. Quality of written communication is assessed throughout the paper and is specifically awarded a maximum of 2 marks in Section B.

Study and revision skills

Students who achieve good grades have good study strategies. Since this is the terminal paper, these strategies should now be well developed. A summary follows:

- Revise topics as you work through the course so that the intensive revision required at exam time is less pressured.
- Be active in your work. Write your own notes using a variety of approaches — a series of bullet points, a flow diagram, an annotated diagram, a spider diagram or a prose account.
- Use a variety of resources so that you develop the level of understanding required to deal with questions that are presented in unfamiliar contexts.
- Organise your notes and keep them up to date.

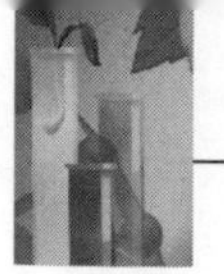

- For the intensive revision in the weeks before the exam make out a revision schedule.
- Revise in a quiet place with no distractions so that you maintain your concentration and produce high quality work.
- Try to keep your final revision active by summarising what you already have.

The examination

You should have gained from the experience of your previous biology exams. However, the following is a list of things for you to remember:

- Prepare the day before by ensuring that you have all the implements that you will need and try to get a good night's sleep.
- With 120 minutes to answer questions worth 90 marks you have some preparation time and should also have some time at the end of the examination to go over your answers.
- Read the questions carefully. There are two aspects to this:
  - Do what the 'command term' asks — for example, if asked to *explain* you should provide reasons for why or how something is happening, not simply *describe* what is happening.
  - Use the information that the examiners provide in the stem to help you answer the question. This information is important since the question may be set in an unfamiliar context.
- The examiners give you guidance about how much you need to write through the number of marks available for each part-question. In Section B, the recommended time is 25 minutes.
- Quality of written communication (QWC) is assessed, so make sure that you pay attention to the way in which you express your ideas and to the use of appropriate biological terminology.

Content Guidance

This section summarises what you need to know and understand for the A2 Unit 2 examination paper. It is divided into eight topics:

- **Respiration** — the biochemistry of aerobic and anaerobic respiration, and the use of different respiratory substrates
- **Photosynthesis** — light harvesting, the light-dependent reactions and the light-independent reactions, and the environmental factors affecting the rate of photosynthesis
- **DNA as the genetic code** — the processes of transcription and translation in the synthesis of proteins and the effects of gene mutation
- **Gene technology** — genetically modified organisms, gene therapy, genome sequencing, knockout gene technology and some of the issues surrounding the use of gene technology
- **Genes and patterns of inheritance** — monohybrid inheritance (dominance, codominance, lethal allelic combinations, multiple alleles and sex linkage), human genetics and pedigrees, and dihybrid inheritance, including gene interaction
- **Population genetics, evolution and speciation** — the use of the Hardy-Weinberg equation in the study of population genetics, the action of selection on genetic variation, evolutionary change and speciation
- **Kingdom Plantae** — the structure and life cycles of mosses, ferns and flowering plants, with a comparison of these plant divisions
- **Kingdom Animalia** — basic structure and feeding in the phyla Cnidaria, Platyhelminthes, Annelida and Chordata, and concluding with evolutionary trends in the kingdom

At various points in the section, there are examiner's tips. These offer guidance on how to avoid difficulties that often occur in examinations.

At the end of each topic there is a list of practical work with which you should be familiar.

Respiration

Energy, adenosine triphosphate and respiration

Energy is defined as 'the capacity to do work'. The work of a cell includes:

- active transport — moving ions and molecules across a membrane against a concentration gradient
- secretion — large molecules produced in some cells are exported by exocytosis
- endocytosis — bulk movement of large molecules and particles into cells
- biosynthesis — anabolic reactions producing large molecules (e.g. proteins from amino acids, cellulose from β-glucose)
- replication of DNA and synthesis of organelles — events during the cell cycle
- contraction of myofibrils — movement of actin filaments over myosin filaments
- activation of molecules — glucose is phosphorylated at the beginning of respiration

The energy for all of this work is made available from **adenosine triphosphate** (ATP). ATP is composed of adenine (a base) attached to a ribose (a pentose sugar) molecule which is attached to a linear sequence of three phosphate groups (see Figure 1).

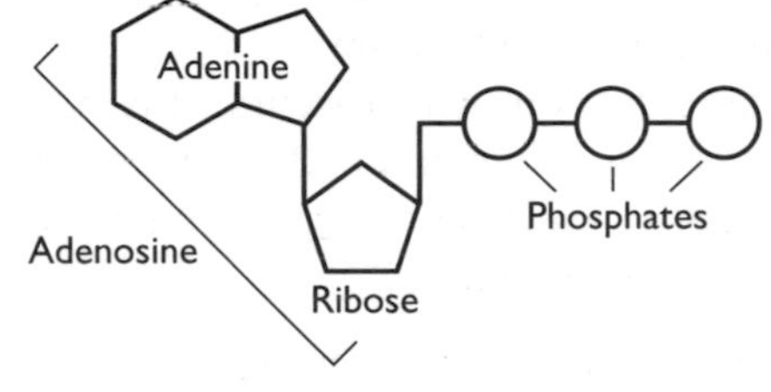

Figure 1 The structure of adenosine triphosphate (ATP)

Energy is released from ATP when it is hydrolysed to ADP and inorganic phosphate (P_i) by the enzyme ATPase. ATP is referred to as 'energy currency' because, like money, it can be used in different contexts, and it is constantly being recycled. At any one time there is only a small pool of ATP in the cell. When ATP is required it is recycled from ADP and P_i by the transfer of energy from the breakdown of glucose, fatty acids and, occasionally, amino acids. These are called **respiratory substrates**. Their breakdown, with the release of energy for ATP synthesis, is called **respiration**. The recycling of ATP is shown in Figure 2.

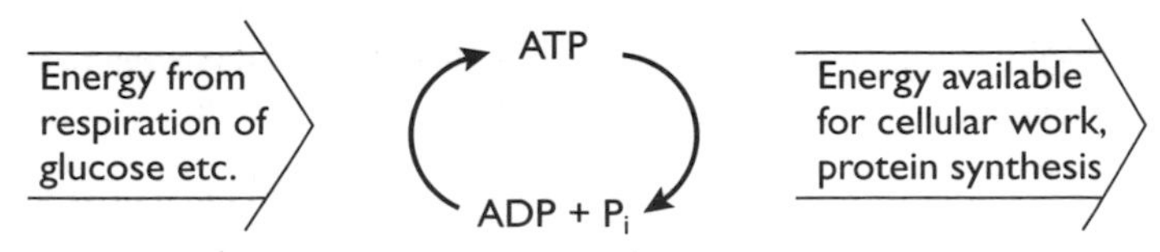

Figure 2 Recycling ATP: its hydrolysis with the release of energy and its synthesis as a result of respiration

Tip Respiration occurs via a series of small steps. Some of these involve the removal of hydrogen (**dehydrogenation**), the removal of carbon dioxide (**decarboxylation**) or the addition of phosphate (**phosphorylation**). You should be familiar with these terms.

During respiration, ATP is produced in two processes:

- **substrate-level phosphorylation** — ATP is produced by the direct transfer of a phosphate group from a phosphorylated substance (a substance with a phosphate group attached) to ADP
- **oxidative phosphorylation** — ATP is produced from ADP and Pi as electrons are transferred along a series of carriers (the electron transport chain)

Two different forms of respiration are recognised:

- **Anaerobic respiration** does *not* require oxygen and can only use glucose (carbohydrate). This is incompletely broken down and so only a little ATP is produced.
- **Aerobic respiration** requires oxygen and uses a variety of respiratory substrates that are completely broken down to produce a great deal of ATP.

Biochemistry of aerobic respiration

Aerobic respiration of glucose occurs in four stages in cells:

1. Glycolysis, which occurs in the cytoplasm.
2. Link reaction (pyruvate oxidation), which occurs in the matrix of the mitochondrion.
3. Krebs cycle, which occurs in the matrix of the mitochondrion.
4. Oxidative phosphorylation (electron transport chain), which occurs across the inner membrane of the mitochondrion.

The **mitochondrion** is the organelle of aerobic respiration.

Glycolysis is the splitting of glucose in a metabolic pathway that has four major steps (see Figure 3). Glycolysis requires glucose, ATP, ADP, P_i and NAD^+. It produces NADH (+ H^+) and ATP.

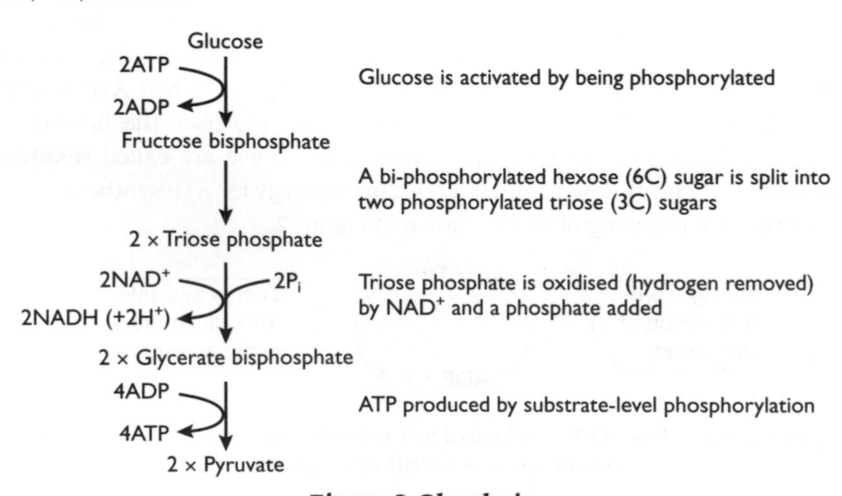

Figure 3 Glycolysis

For every molecule of glucose used in glycolysis, two molecules of ATP are used initially. However, four molecules of ATP are produced, so there is a net production of two ATP molecules per glucose molecule (this occurs in both aerobic and anaerobic respiration). During aerobic respiration, additional energy is available from reduced NAD (two molecules of reduced NAD are produced per glucose). For this additional energy to be made available, electrons from reduced NAD must pass into the mitochondrion (by a 'shuttle' system) to produce more ATP by oxidative phosphorylation. There are two 'shuttle' systems, one of which operates at the cost of one molecule of ATP. Pyruvate is transported into the mitochondrion where, if oxygen is available, it is further metabolised. Two molecules of pyruvate are produced per molecule of glucose. Therefore, when considering the breakdown of one molecule of glucose, the products of this further metabolism of pyruvate must be doubled.

In the mitochondrion, pyruvate is initially broken down in a **link reaction** (so called because it links with the Krebs cycle). The link reaction involves the removal of hydrogen (dehydrogenation) and the removal of carbon dioxide (decarboxylation) from pyruvate, with the formation of an acetyl group (see Figure 4).

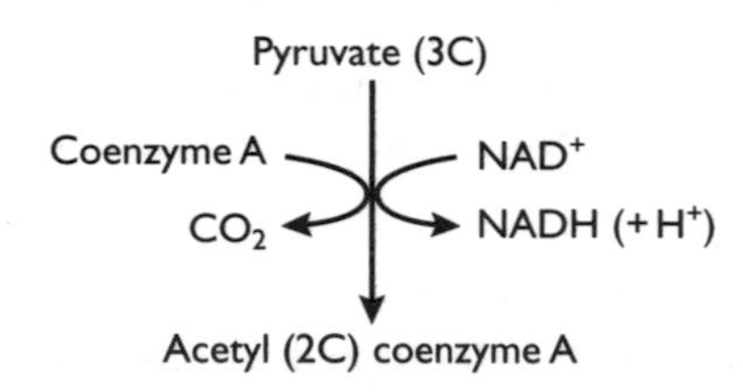

Figure 4 The link reaction (oxidation of pyruvate)

The carbon dioxide diffuses out of the mitochondrion and out of the cell. The hydrogen that is removed is picked up by NAD^+ producing NADH (reduced NAD). The two-carbon acetyl group is carried by coenzyme A to produce acetyl coenzyme A.

Coenzymes work closely with the enzymes of respiration, carrying products of the breakdown to be used elsewhere:

- NAD^+ carries hydrogen, as NADH (+ H^+), to be used in the electron transport chain (oxidative phosphorylation)
- coenzyme A carries the acetyl group, as acetyl coenzyme A, to be used in the Krebs cycle

In the **Krebs cycle** the acetyl group, from acetyl coenzyme A, combines with a four-carbon acid to form a six-carbon acid. During each cycle:

- two steps involve the removal of carbon dioxide molecules, which diffuse out of the mitochondrion and out of the cell
- one step is a substrate-level phosphorylation, producing an ATP molecule
- four steps involve dehydrogenations, i.e. hydrogen is removed from the substrate

The Krebs cycle is shown in Figure 5.

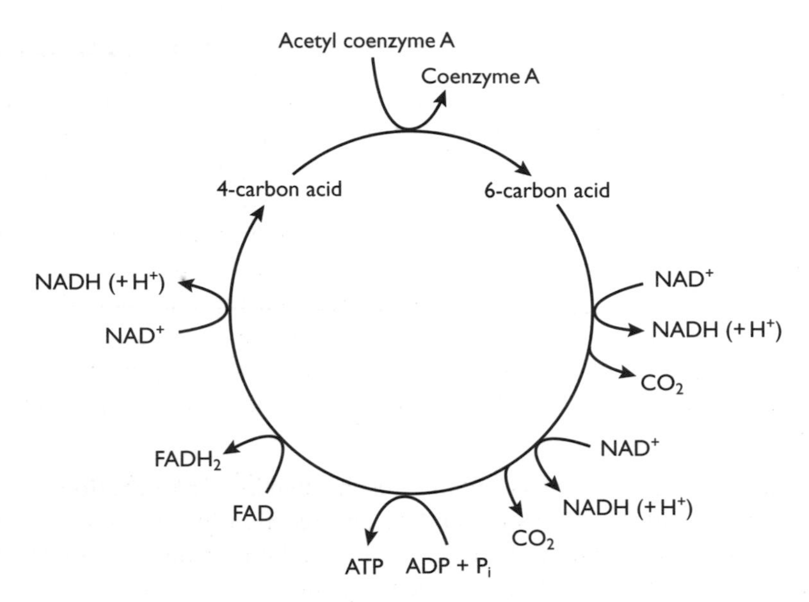

Figure 5 Krebs cycle

The removal of hydrogen is the most significant feature of the Krebs cycle. In three instances hydrogen is carried by NAD^+ and in one case hydrogen is carried by FAD, which is another coenzyme.

The hydrogens removed during glycolysis, the link reaction and the Krebs cycle are the source of electrons for the **electron transport chain**. In this chain, ATP is produced through **oxidative phosphorylation**. The process has the following features:

- Hydrogens are passed along initially and then subsequently only their electrons. (Remember that a hydrogen atom consists of a proton, or hydrogen ion, and an electron.)
- The hydrogens, or their electrons, are passed through a series of carriers. NAD^+, flavoprotein and coenzyme Q act as hydrogen carriers; cytochromes and cytochrome oxidase are electron carriers.
- As a carrier receives hydrogens, or electrons, it is reduced. As it passes the hydrogens or electrons along it is oxidised. (This alternate reduction and oxidation of a carrier is called a redox reaction.)
- Oxygen acts as the final electron acceptor. It is reduced, forming water.
- The carriers lie at successively lower energy levels so that as the hydrogen ions and electrons are transferred energy becomes available.
- Sufficient energy becomes available at certain points in the chain to produce ATP through oxidative phosphorylation.
- From each NADH sufficient energy is released to produce three ATP molecules. Two ATP molecules are produced from each $FADH_2$ (since hydrogen enters the chain from $FADH_2$ at a point after the production of the first ATP). Therefore NADH has a greater yield of ATP.

- Electron transfer and oxidative phosphorylation are tightly coupled: one process only takes place if the other process can take place. For example, if ADP and P_i are not available for phosphorylation, then no electron transfer can take place and the coenzymes are all reduced, no matter how much oxygen is available.

The electron transport chain and oxidative phosphorylation are shown in Figure 6.

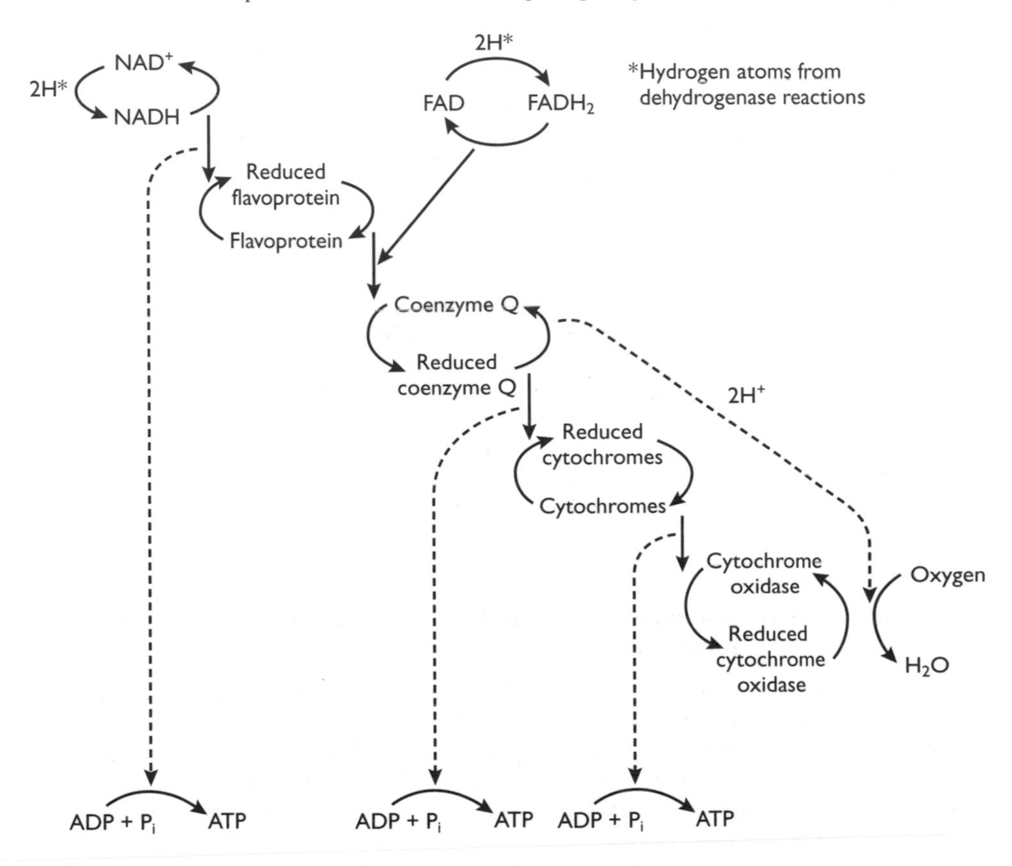

Figure 6 The electron transport chain and oxidative phosphorylation

Tip Figure 6 indicates that 3 ATP molecules are produced per NADH and 2 are produced per $FADH_2$. However, recent research suggests that these figures are 2.5 and 1.5 ATP molecules respectively. Since both sets of figures are used in textbooks, either set is acceptable in examinations.

The total number of ATP molecules produced from the aerobic breakdown of glucose is shown in Figure 7, which also serves as a useful overview of the process of aerobic respiration. Note that glycolysis takes place in the cytoplasm, the link reaction and Krebs cycle take place in the mitochondrial matrix, and electron transport and oxidative phosphorylation take place in the inner membrane of the mitochondrial envelope.

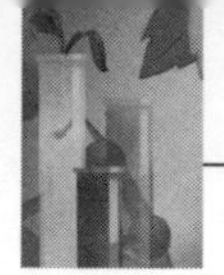

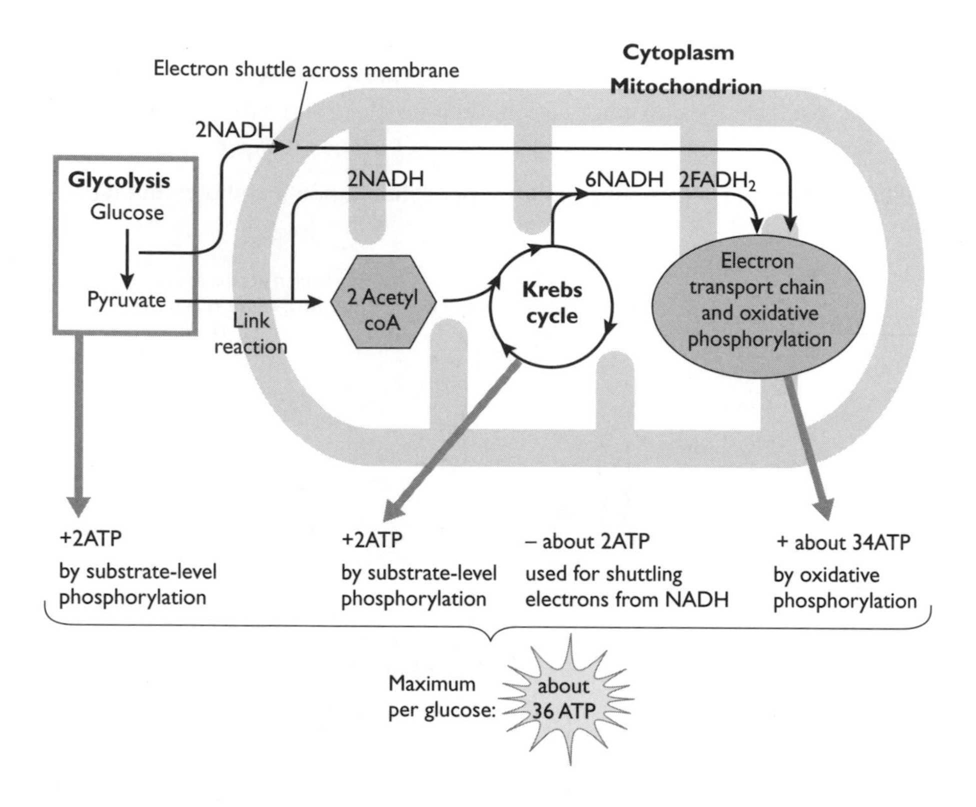

Figure 7 The total number of ATP molecules produced from the aerobic breakdown of one molecule of glucose

The figure quoted for the total number of ATP molecules produced from one molecule of glucose in aerobic respiration varies according to a number of factors: whether the ratio of ATP produced from reduced NAD and reduced FAD is taken as 3:2 or 2.5:1.5 and whether account has been taken of the usage of ATP to shuttle electrons from the reduced NAD produced in the cytoplasm into the mitochondrion. Figure 7 is based on three ATP molecules produced from each reduced NAD molecule, two ATP molecules being produced from each reduced FAD and one ATP being expended to shuttle electrons into the mitochondrion.

Other respiratory substrates

Respiratory substrates other than glucose can be respired aerobically.

Other **carbohydrates** are funnelled through the glycolytic pathway:

- sugars such as fructose enter glycolysis
- polysaccharides such as starch (in plants) and glycogen (in animals) are hydrolysed into glucose or glucose phosphate and so enter glycolysis

Triglycerides (lipids such as fats and oils) are hydrolysed into glycerol and fatty acids. Glycerol enters the glycolytic pathway. Fatty acids are energy rich, and ATP is generated in the mitochondrion either from initial splitting into two-carbon fragments (β-oxidation) or from the metabolism of the acetyl coenzyme A formed.

Proteins are hydrolysed to amino acids, which can be used to supply energy in dire situations such as starvation. The amino group is removed (deamination) and the remnant organic acid is converted into pyruvate or acetyl coenzyme A (mostly) or several of the acids in Krebs cycle. (Excess dietary amino acids are converted to carbohydrates or triglycerides.)

All substrates respired aerobically are funnelled through the Krebs cycle, often via acetyl coenzyme A, which is referred to as the focal point of respiratory metabolism. This is illustrated in Figure 8.

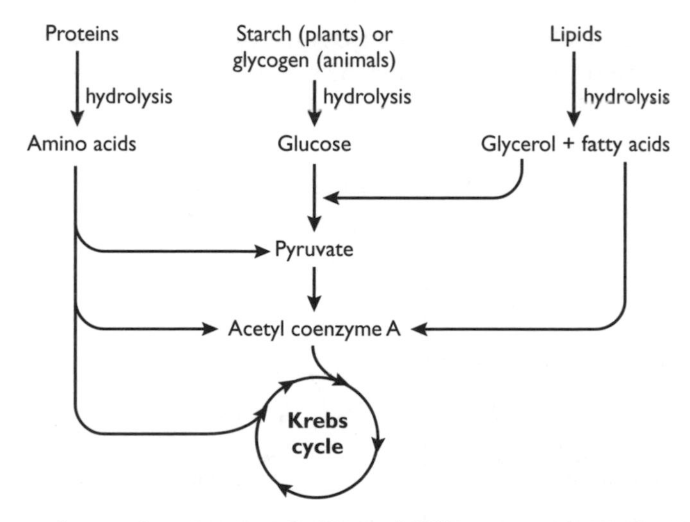

Figure 8 The respiratory metabolism of different respiratory substrates

The energy values of respiratory substrates are shown in Table 1. Since 1 gram of lipid yields more than twice as much energy as 1 gram of carbohydrate, only half the mass is needed to store an equivalent amount of energy. Organisms that have limited opportunities for obtaining food generally carry their 'energy store' as lipid. Examples include the fats stored in camels' humps and in migrating birds such as ducks and the stores of oils in numerous types of plant seed.

Table 1 Energy values of respiratory substrates

Respiratory substrate	Energy yield/kJ g^{-1}
Carbohydrate (polysaccharides and sugars)	16
Lipids (triglycerides)	39
Proteins (amino acids)	17

However, lipids can only be respired aerobically.

Biochemistry of anaerobic respiration

When the supply of oxygen is limited, the rate at which hydrogen carriers are oxidised (e.g. to NAD^+ and FAD) is restricted and the cell may resort to anaerobic respiration. If there is no oxygen available, there is no terminal electron acceptor so all the carriers in the hydrogen/electron transport chain remain reduced and the link reaction and Krebs cycle cannot take place.

Anaerobic respiration is essentially glycolysis, with additional reactions to regenerate NAD^+. The coenzyme NAD^+ is required for the oxidation of triose phosphate to glycerate bisphosphate (see Figure 3). In this reaction the NAD^+ is converted into NADH. Shuttling its electrons into the mitochondrion is limited by the availability of oxygen. Plants (and fungi) and animals have developed different pathways for regenerating NAD^+ without the need for oxygen (see Figure 9). In both pathways, two molecules of ATP are produced from the metabolism of one molecule of glucose.

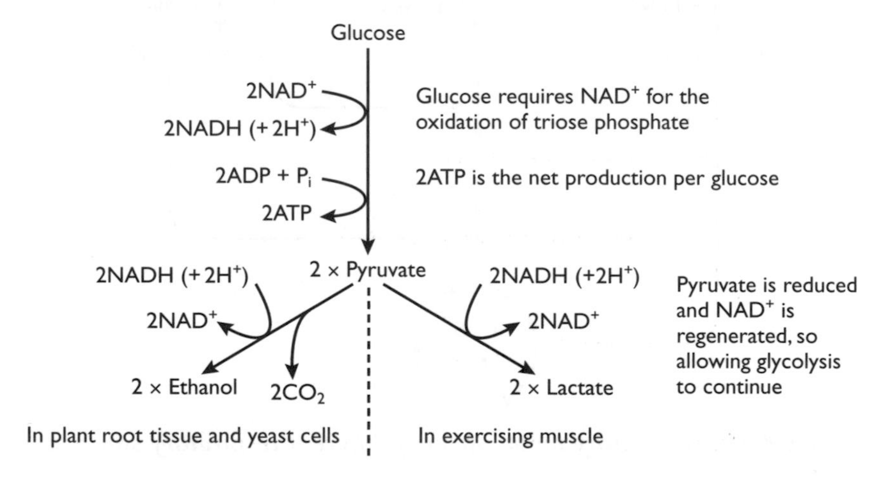

Figure 9 The anaerobic pathways of different organisms

The different pathways used by some plants (and fungi) and animals reflect different strategies:

- In plants and fungi, the anaerobic pathway allows the organism (e.g. yeast in suspension) or tissue (e.g. plant root in waterlogged conditions) to survive for a period of time without a source of oxygen. The toxic effect of the waste products is reduced because they diffuse readily out of the organism — CO_2 is a one-carbon molecule and ethanol is a two-carbon molecule.
- In animals, the anaerobic pathway allows some tissues (e.g. mammalian muscle) to generate extra ATP over and above that generated aerobically. While only two molecules of ATP are produced per glucose in anaerobic respiration, this can occur extremely rapidly. It is this extra energy release that allows us to undertake strenuous exercise — for example, sprinting. The limiting factor is our tolerance to lactate, which is toxic, so strenuous exercise can only take place for a brief period. After strenuous activity, aerobic respiration is heightened because extra oxygen is

supplied as we breathe rapidly and the heart rate increases. The extra ATP produced is used to convert lactate back to glucose — the 'oxygen debt' is paid back.

Respiratory quotients

The amount of carbon dioxide produced relative to the amount of oxygen consumed provides information about:

- the type of respiration taking place (i.e. aerobic or anaerobic)
- the nature of the respiratory substrate (e.g. carbohydrate or lipid)

The relative amount of carbon dioxide produced is called the **respiratory quotient** (RQ).

$$RQ = \frac{CO_2 \text{ produced}}{O_2 \text{ consumed}}$$

If the RQ is greater than 1, then extra carbon dioxide is being produced. This means that both aerobic and anaerobic respiration are taking place in, for example, yeast or plant tissue. (In yeast, if only anaerobic respiration is taking place, then theoretically the RQ value would be infinite. Remember too that anaerobic respiration in animal tissue does not produce carbon dioxide and that anaerobic respiration can use only carbohydrate as the respiratory substrate.)

If the RQ is less than or equal to 1, then aerobic respiration only is taking place, at least in yeast or plant tissue:

- If the RQ is 1, the respiratory substrate is carbohydrate.
- If the RQ is approximately 0.7, the respiratory substrate is lipid (fat or oil).
- If the RQ is approximately 0.9, the respiratory substrate is protein.

In general, however, a variety of respiratory substrates (mostly a combination of carbohydrate and lipid) are used. This is why, in humans, the RQ is in the region of 0.85.

Practical work

Refer to the use of a simple respirometer as in AS Unit 2, part 2.1 (b) of the specification.

- Calculation of RQ values.

Demonstrate the role of hydrogen acceptors using a redox indicator (e.g. methylene blue or tetrazolium chloride).

Photosynthesis

Photosynthesis is the process by which green plants, using the energy from sunlight, produce a simple carbohydrate from the inorganic molecules carbon dioxide and water. Photosynthesis also releases oxygen, which is needed for aerobic respiration.

The simple carbohydrate synthesised is used to produce other carbohydrates (such as starch, sucrose and cellulose), fatty acids and lipids, and amino acids and proteins. In doing this, plants make their own food — they are **autotrophic**.

Links Green plants have numerous adaptations for photosynthesis. Many of these relate to the structure of the leaf, which is covered in the AS Unit 1 and AS Unit 2 guides in this series, to which you should refer. Others relate to the structure of the chloroplast, which is covered in the AS Unit 1 guide.

The structure of a chloroplast is shown in Figure 10.

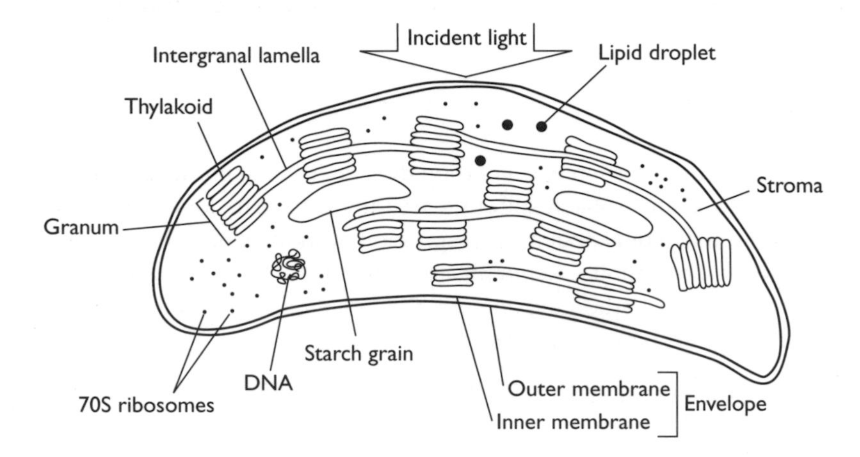

Figure 10 The structure of a chloroplast

Chloroplasts are orientated within the cell to ensure maximum exposure of the thylakoid surface to the direction of incident light.

Light is absorbed by a number of different pigments on the lamellae (e.g. thylakoids). There are two main classes of photosynthetic pigments:

- **chlorophylls** — chlorophyll *a* and chlorophyll *b*
- **carotenoids** — for example, β-carotene, the pigment in carrots

The **absorption spectrum** (Figure 11 (a)) is a graph that shows how much light a particular pigment absorbs at each wavelength. The chlorophylls absorb light in the blue–violet and red parts of the spectrum. The carotenoids absorb light from the blue–violet part of the spectrum. Different pigments absorb light of different wavelengths so the combined effect is to increase the range over which light is effectively absorbed (see Figure 11 (a)). The absorption spectrum can be compared with a graph of the rate of photosynthesis against wavelength, which is called an **action spectrum** and is shown in Figure 11 (b). The shape of the action spectrum is similar to that of the absorption spectrum for the combined photosynthetic pigments — for example, the absorption spectrum shows that little green light is absorbed and therefore that green light is relatively ineffective in photosynthesis.

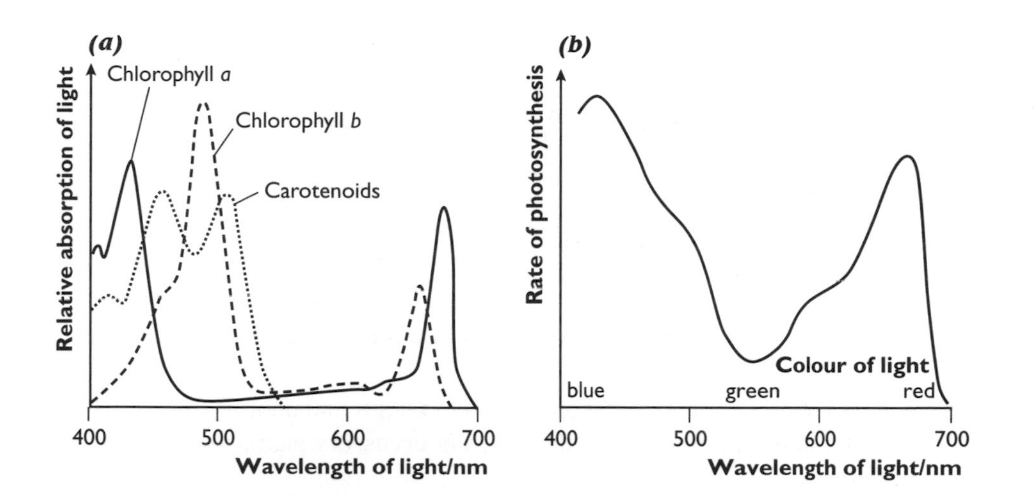

Figure 11 (a) The absorption spectrum for photosynthetic pigments (b) The action spectrum for photosynthesis

There are three main stages to photosynthesis:

- Light harvesting — the capturing of light by groups of photosynthetic pigments called photosystems. The light energy is used to raise the energy levels of electrons.
- Light-dependent reactions — energised electrons from the photosystems are used to produce the energy-rich compounds ATP and NADPH.
- Light-independent reactions — carbon dioxide is 'fixed' (incorporated) into organic form and the products of the light-dependent stage, ATP and NADPH, are used to convert this to simple carbohydrate.

Light harvesting

The absorption of light is carried out by clusters of pigment molecules organised into photosynthetic units or **photosystems** located on the lamellae of chloroplasts. Each unit contains several hundred chlorophyll and carotenoid molecules. One particular chlorophyll *a* molecule, called a **primary pigment**, acts as a **reaction centre** for the photosystem. The remaining pigment molecules of the photosystem are called **accessory pigments** and represent an **antenna complex** absorbing light energy. Light energy absorbed by an accessory pigment creates an excitation energy that is passed along a chain of pigment molecules to the reaction centre. Resonance transfer of energy occurs, i.e. it is the energy that is transferred not the electrons. Energy from many pigment molecules in the antenna complex is funnelled to the reaction centre. Energy reaching the reaction centre causes electrons in the primary chlorophyll *a* molecule to move up to a higher energy level. Such is the increased energy level of the electron that it is emitted by the chlorophyll *a* and taken up by an electron acceptor. Light harvesting is illustrated in Figure 12.

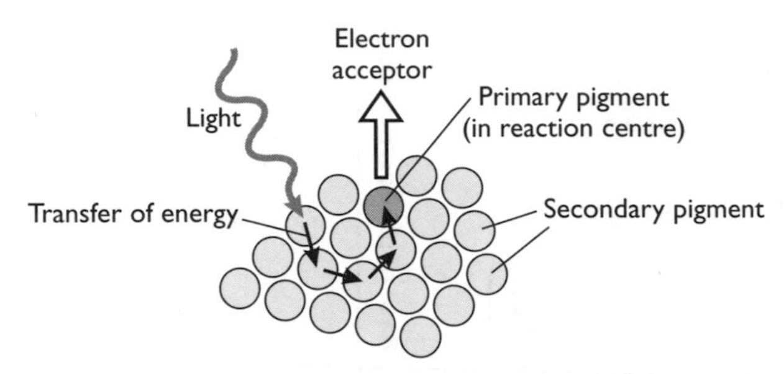

Figure 12 Light harvesting in a photosystem

There are two photosystems, called photosystem I (PS I) and photosystem II (PS II):

- They differ in the proportions of the different accessory pigments that make up the cluster.
- They differ in their primary pigments. PS I has a form of chlorophyll *a* that has a light absorption peak at 700 nm and is called P700. PS II has a form of chlorophyll *a* that has a light absorption peak at 680 nm and is called P680.

Light-dependent reactions

The excitation of the electrons that follows the harvesting of light by photosystems II and I results in an electron transfer pathway described as a 'Z-scheme'. The sequence of events is shown in Figure 13.

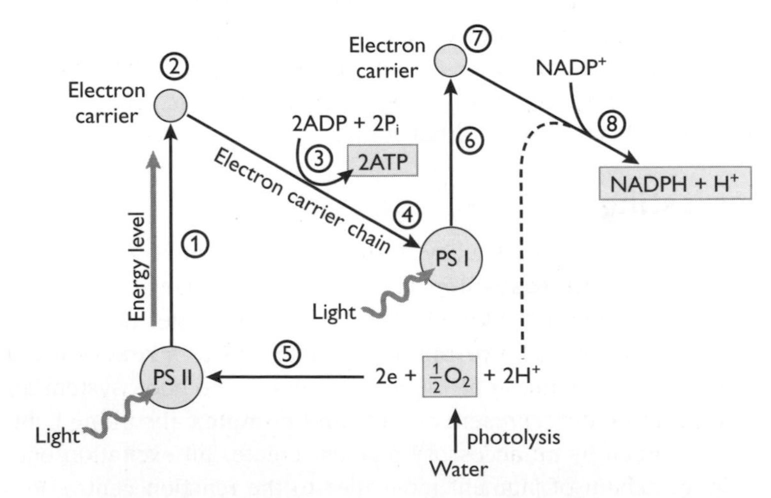

Figure 13 The light-dependent reactions of photosynthesis

(1) Light energy is trapped in PS II and boosts electrons to a higher energy level.
(2) The electrons are received by an electron acceptor.
(3) The electrons are passed from the electron acceptor along a series of electron carriers (cytochromes). Electron transfer is coupled to the **synthesis of ATP** (**photophosphorylation**).
(4) The electrons replace those lost in PS I.

(5) Electrons lost by PS II are removed from water molecules, the resultant **photolysis** of which produces **oxygen** and hydrogen ions.
(6) Light energy trapped by PS I boosts the electrons to an even higher energy level.
(7) The electrons are removed by an electron acceptor.
(8) The hydrogen ions from the water combine with the electrons from the second electron acceptor and these **reduce $NADP^+$ to NADPH** (+ H^+).

The most important consequences of this electron transfer are:

- the synthesis of ATP via photophosphorylation and the production of NADPH, both of which are used in the metabolic reactions whereby simple carbohydrate is synthesised
- the release of oxygen as a waste product and its diffusion out of the chloroplast; any that is not used in respiration diffuses out of the cell and out of the leaf

Tip Both NAD^+ and $NADP^+$ act as coenzymes that carry hydrogen. You must not confuse them. Remember that NAD^+ is used in respiration and NAD**P**$^+$ is used in **p**hotosynthesis.

The electron acceptors and carriers involved in the light-dependent reactions are closely associated with the photosystems and are located on the granal lamellae (thylakoids) of the chloroplasts.

Light-independent reactions: the Calvin cycle

The light-independent reactions (Calvin cycle) take place in the stroma of the chloroplasts. Four stages are recognised:

- **Fixation of carbon dioxide** — carbon dioxide enters the stroma and combines with the five-carbon compound **ribulose bisphosphate** (**RuBP**). This reaction is catalysed by the enzyme **ribulose bisphosphate carboxylase** (**Rubisco**). Two molecules of glycerate phosphate (GP) are formed.
- **Reduction of glycerate phosphate** — ATP and NADPH from the light-dependent reactions convert GP into **triose phosphate** (**TP**). ATP provides the energy to drive this reaction and NADPH provides the reducing power for the reaction. The ADP and $NADP^+$ formed leave the stroma and enter the granal lamellae.
- **Regeneration of ribulose bisphosphate** — five out of every six TP molecules produced are used to regenerate ribulose bisphosphate using the remainder of the ATP from the light-dependent reactions as a source of phosphate and energy.
- **Product synthesis** — while, for every six TP molecules produced five are used to regenerate RuBP, one TP represents 'profit'. The TP molecules so generated are used to produce a range of products — for example, hexose sugar is produced from two TP molecules. Other molecules are produced, including starch (storage carbohydrate from the hexose sugar), glycerol and fatty acids (and therefore lipids), and amino acids (and therefore proteins) with nitrogen being obtained from absorbed nitrate ions.

The light-independent reactions are shown in Figure 14.

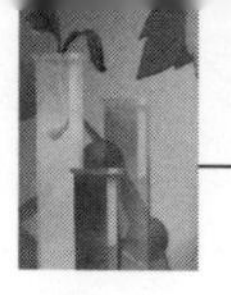

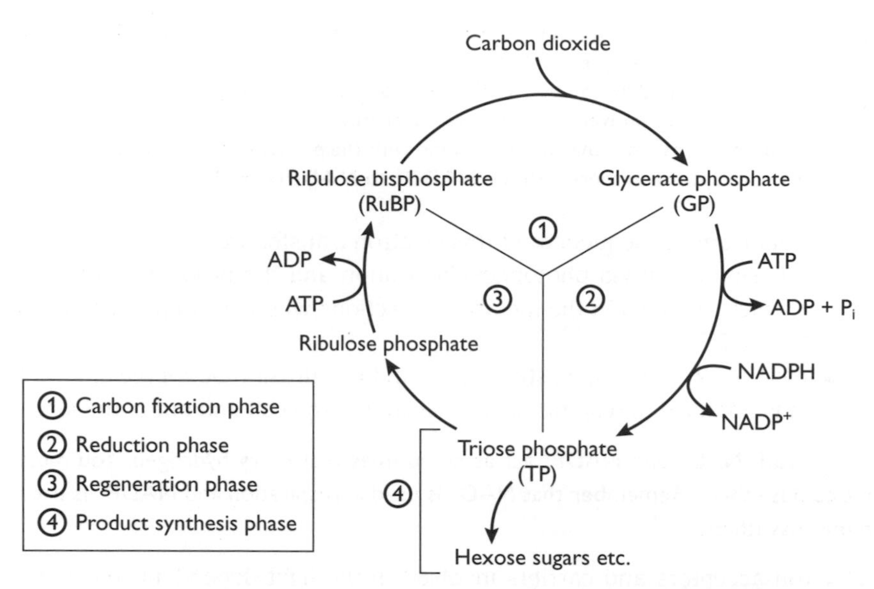

Figure 14 The light-independent reactions: the Calvin cycle

Environmental factors that affect the rate of photosynthesis

The main factors that affect the rate of photosynthesis are light intensity, carbon dioxide concentration and temperature. These are called **limiting factors**. A limiting factor is that factor, of a number of possible factors, which is determining the rate at which the process is taking place. When a limiting factor is increased, the process takes place at a faster rate. Given that there is some carbon dioxide available and some warmth, light intensity is a limiting factor at low light intensities. This is illustrated in Figure 15.

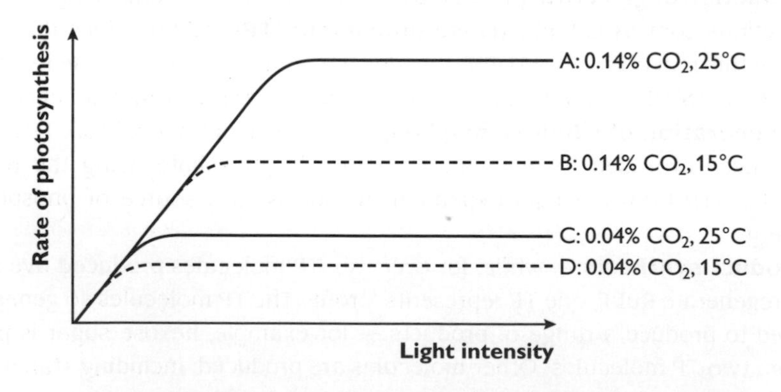

Figure 15 The effects of light intensity, carbon dioxide concentration and temperature on the rate of photosynthesis

Figure 15 is a complex graph showing the effects of light intensity, carbon dioxide concentration and temperature on the rate of photosynthesis. Light is not the limiting factor at higher light intensities (the plateau region of the graph). As light intensity is increased, the rate of photosynthesis is eventually limited by some other factor.

Carbon dioxide is the limiting factor at high light intensity and over a range of temperatures:

- At 15°C, the rate of photosynthesis increases significantly from D to B.
- At 25°C, there is an even greater increase from C to A.

Temperature appears to be an important limiting factor at high light intensity *and* at the higher carbon dioxide concentration. There is little increase in the rate of photosynthesis at the lower 0.04% carbon dioxide concentration (from D to C). At 0.14% carbon dioxide, the rate of photosynthesis increases significantly from B to A.

The effects of the three factors are shown in Table 2.

Table 2 The effects of environmental factors on the rate of photosynthesis

Environmental factor	Reason for influence	Effect of factor	Effect in the environment
Light intensity	Greater light intensity increases the energy available for the light-dependent reactions and, therefore, the production of ATP and NADPH for the light-independent reactions	As light intensity increases, the rate of photosynthesis increases proportionally, until it is limited by some other factor	On a winter's day the light intensity may be limiting (if it is not too cold); short day-length, during which photosynthesis is possible, is a major influence
Carbon dioxide concentration	An increase in CO_2 concentration increases the carboxylation of RuBP and, therefore, the production of GP	An increase in carbon dioxide concentration increases the rate of photosynthesis, until it is limited by another factor	In bright conditions during the summer, CO_2 is probably limiting; levels of CO_2 may be increased artificially in a greenhouse; the global increase in CO_2 should increase the rate of photosynthesis
Temperature	Enzymes catalyse the light-independent reactions; as the enzymes approach their optimum temperature, activity increases; above the optimum temperature enzymes are denatured	Temperature is only a limiting factor at very low or very high temperatures or when there are relatively high levels of light and CO_2	On a very cold winter's day, temperature may be limiting; in a greenhouse, additional heating may be provided during the winter; a temperature of around 25°C is optimal; plants survive high summer temperatures because transpiration has a cooling effect

Links The rate of photosynthesis is measured as either the amount of carbon dioxide taken up or oxygen produced, although these are both influenced by the rate of respiration as well as the rate of photosynthesis. When both rates are equal, a compensation point is achieved (refer to the AS Unit 2 guide pp. 19–20 in this series).

There are also links with an understanding of gross production and net production (gross production minus respiration). Refer to the A2 Unit 1 guide in this series.

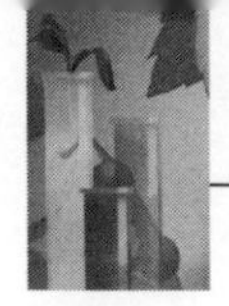

Practical work

Refer to the use of the Audus apparatus as in AS Unit 2, part 2.1 (b) of the specification.

- The effect of light intensity and carbon dioxide concentration on the rate of photosynthesis.

Carry out paper chromatography of plant pigments.

- Preparation and running of the chromatogram.
- Calculation of R_f values.

Demonstrate the role of hydrogen acceptors using a redox indicator (such as DCPIP).

DNA as the genetic code

A **gene** is a length of DNA that acts as a coding device for the synthesis of a particular protein. More precisely, it is a sequence of nucleotide bases which determines the order of the amino acids in the protein produced.

Links The structure of nucleotides and nucleic acids (DNA and RNA), polypeptides and proteins is covered in the guide to AS Unit 1 in this series.

It is the sequence of nucleotide bases that represents the **genetic code**. The main features of the genetic code are as follows:

- It is a **three-base (triplet) code**. A sequence of three nucleotide bases codes for an amino acid. There are four bases arranged in groups of three, so the number of different sequences is 4^3 or 64. As there are only 20 amino acids used for protein synthesis, this is more than enough sequences. (Note that a one-base code would only code for four amino acids and a two-base code would only code for 4^2 or 16 amino acids.)
- It is a **degenerate code**. Most amino acids have more than one triplet code, since there are more triplet codes (64) than amino acids (20). For example, the amino acid leucine has six codes and many have four (see Table 3, p.27).
- It is a **non-overlapping code**. The triplet codes are read separately, i.e. each base is read only once. For example, AGAGCG is read as AGA for one amino acid and GCG for the next, rather than AGA, GAG, AGC, GCG.

DNA does not act directly in protein synthesis. It is transcribed into a single-stranded RNA molecule that acts as a messenger. There are a number of reasons for this:

- DNA is a very long molecule (so long, that in eukaryotes, each is supercoiled into a chromosome during nuclear division). It is too long to move out of the nucleus into the cytoplasm where protein synthesis takes place.
- DNA is retained in the nucleus where, as the molecule of inheritance, it is better protected from the possibility of damaging changes.

- DNA is used to produce multiple copies of messenger RNA and so many proteins can be synthesised simultaneously.

The result is that there are two stages to protein synthesis:
- The genetic code in the DNA molecule in the nucleus is first copied into a molecule of messenger RNA (mRNA) in the process of **transcription**.
- The mRNA molecule then moves out of the nucleus onto a ribosome where its code is used to direct the synthesis of a polypeptide in the process of **translation**.

Protein synthesis: transcription

An overview of protein synthesis is given in Figure 16.

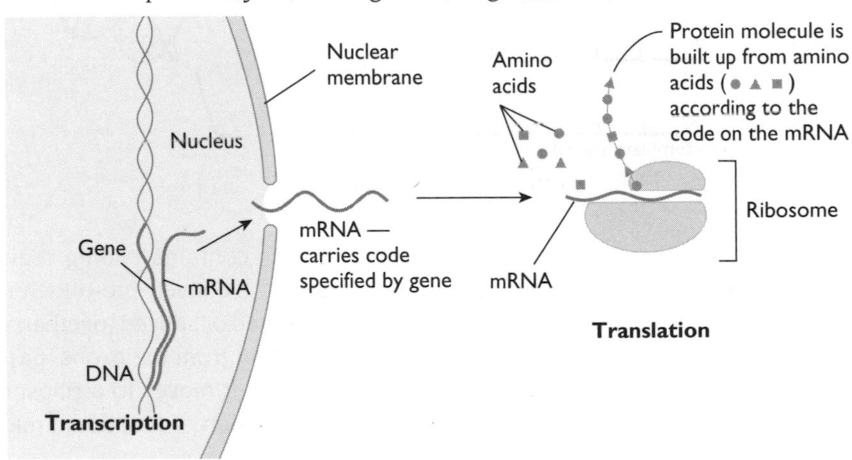

Figure 16 An overview of protein synthesis

The process of transcription (Figure 17) goes through the following sequence:
- The two strands of DNA are separated by the enzyme **helicase** using energy from ATP.
- The enzyme **RNA polymerase** binds to a region of the DNA near the beginning of the gene that is going to be transcribed.
- One of the exposed strands — the **template strand** — is used as a template for mRNA production. As RNA polymerase moves along the template strand, ribonucleotides are taken from the nucleoplasm and matched up by a process of complementary base pairing.
- **Complementary base pairing** involves the bases in the ribonucleotides forming hydrogen bonds with the exposed bases on the DNA template strand, according to the base pairing rules: a ribonucleotide with U binds with A in the DNA strand, G with C, A with T and C with G. The RNA polymerase binds the newly arrived ribonucleotides along their sugar-phosphate backbone.
- As the RNA polymerase moves along, the DNA double helix reforms. When the enzyme reaches the end of the gene it releases the fully formed **mRNA**. The mRNA produced is complementary to the nucleotide base sequence on the template strand of the DNA.

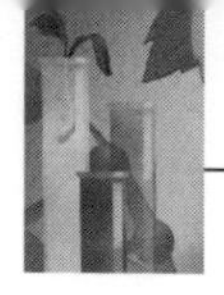

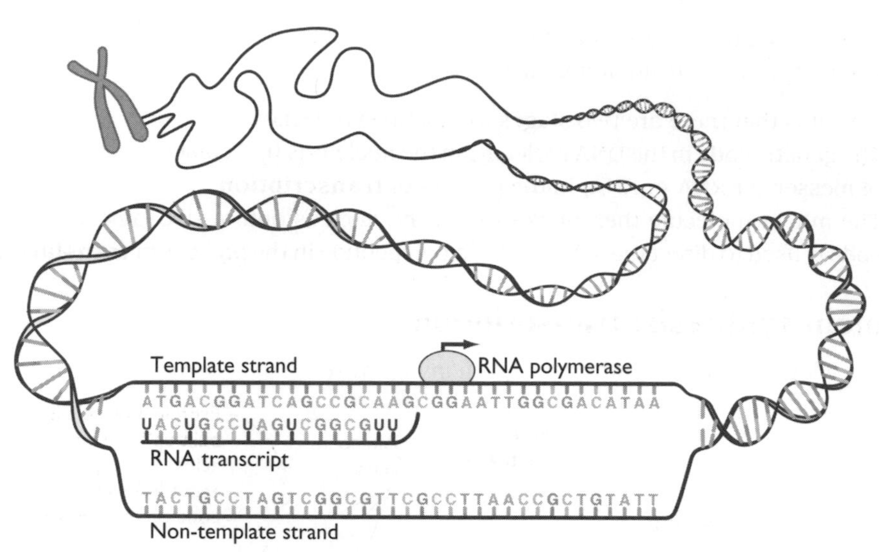

Figure 17 Transcription

In eukaryotic cells, the first mRNA formed (pre-mRNA) contains coding regions (**exons**) interspersed with non-coding regions (**introns**). To convert pre-mRNA into mRNA, the introns are cut out and the remaining exons joined or spliced together. The final version of mRNA, which now contains only the coding from the exons, passes out of the nucleus through a pore in the nuclear envelope. It moves to a ribosome, where the synthesis of the protein takes place. The production of functional mRNA is shown in Figure 18.

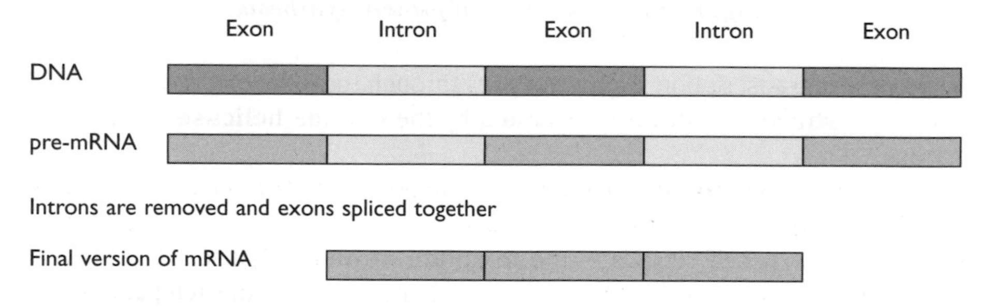

Figure 18 Production of functional mRNA — cutting out introns and splicing together the remaining exons

Protein synthesis: translation

The components of translation are:

- mRNA, containing a sequence of bases, triplets of which are called **codons**
- ribosomes, each of which consists of two subunits, one small and one large
- transfer RNA (tRNA) molecules, each of which carries a specific amino acid at one end and a triplet base code, called an **anticodon**, at the other

The mRNA genetic codes (codons) for amino acids are shown in Table 3. The different amino acids are represented by three-letter abbreviations.

Table 3 The mRNA genetic codes (codons) for amino acids

First position (5′ end)	Second position: U	Second position: C	Second position: A	Second position: G	Third position (3′ end)
U	UUU Phe	UCU Ser	UAU Tyr	UGU Cys	U
U	UUC Phe	UCC Ser	UAC Tyr	UGC Cys	C
U	UUA Leu	UCA Ser	UAA stop	UGA stop	A
U	UUG Leu	UCG Ser	UAG stop	UGG Trp	G
C	CUU Leu	CCU Pro	CAU His	CGU Arg	U
C	CUC Leu	CCC Pro	CAC His	CGC Arg	C
C	CUA Leu	CCA Pro	CAA Gln	CGA Arg	A
C	CUG Leu	CCG Pro	CAG Gln	CGG Arg	G
A	AUU Ile	ACU Thr	AAU Asn	AGU Ser	U
A	AUC Ile	ACC Thr	AAC Asn	AGC Ser	C
A	AUA Ile	ACA Thr	AAA Lys	AGA Arg	A
A	AUG Met/start	ACG Thr	AAG Lys	AGG Arg	G
G	GUU Val	GCU Ala	GAU Asp	GGU Gly	U
G	GUC Val	GCC Ala	GAC Asp	GGC Gly	C
G	GUA Val	GCA Ala	GAA Glu	GGA Gly	A
G	GUG Val	GCG Ala	GAG Glu	GGG Gly	G

The process of translation (Figure 19) follows the following sequence:

- The first two codons of the mRNA enter a ribosome, the first at the peptidyl (P) site and the second at the aminoacyl (A) site. The first codon is a start (initiation) codon.
- Transfer RNA molecules (with specific amino acids attached) that have complementary anticodons to the first two codons of the mRNA bind to those codons.
- A peptide bond, requiring energy from an ATP molecule, is formed between the amino acids carried by these two tRNA molecules.
- The ribosome moves along the mRNA by one codon, bringing the third codon onto the aminoacyl site on the ribosome. The tRNA that is freed returns to the cytoplasm where it will pick up another amino acid.
- A tRNA with a complementary anticodon binds with the third mRNA codon, bringing its amino acid into position next to the second amino acid held at the peptidyl site. The amino acids bond and the chain is lengthened.
- The ribosome moves along the mRNA by one codon, bringing a further codon onto the aminoacyl site. An amino acid carrying tRNA with a complementary anticodon binds with the codon and the amino acid bonds with the existing chain of amino acids held at the peptidyl site.

- The process is repeated until a stop or termination code is in position. This causes the polypeptide to be released. Translation of a polypeptide is complete.

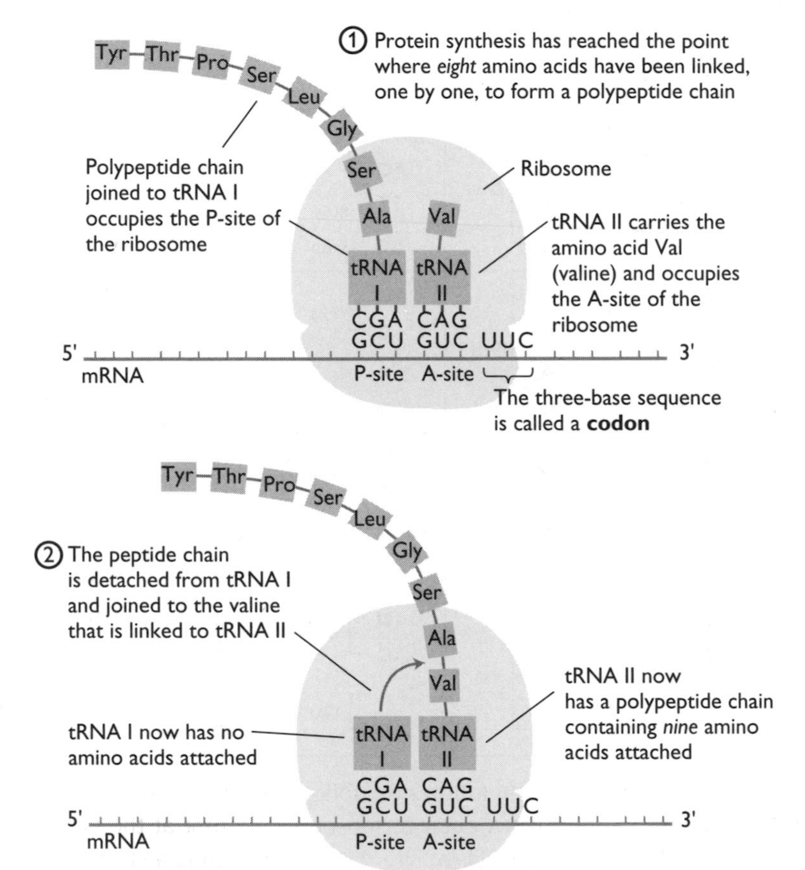

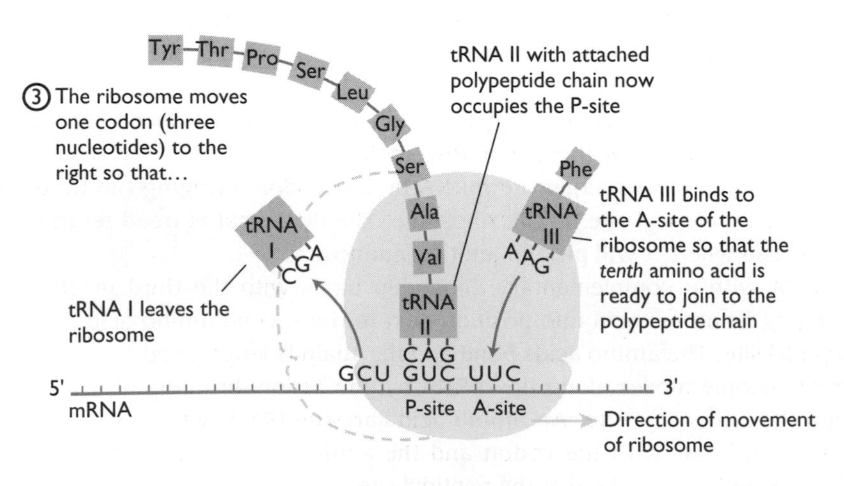

Figure 19 Translation

In eukaryotic cells, the ribosomes are usually associated with endoplasmic reticulum (ER), forming rough ER. Newly synthesised polypeptides enter the ER and vesicles containing the polypeptides are 'nipped' off and move to the Golgi apparatus. Here they fuse with the formative face. In the Golgi apparatus the polypeptides are processed to produce final functional proteins.

These proteins include cell surface proteins, channel proteins, antibodies, some hormones (e.g. insulin) and enzymes. Enzymes are involved in the control of all metabolic pathways and thus in the synthesis of all non-protein molecules found in cells. Whatever their role, proteins determine the characteristics of an organism.

Gene mutation

A **gene mutation** is a spontaneous change in the DNA molecule. It occurs most often as a result of a mistake when DNA is replicating. Two examples are nucleotide base substitutions and deletions:

- In a substitution, one base is replaced by another base.
- In a deletion, one base is omitted altogether.

In the following **base substitution**, a thymine on the template strand of DNA is replaced (mistakenly) by an adenine. The result is a change in the DNA code, a change in the mRNA codon for the amino acid brought onto the ribosome and so in the amino acid sequence of the polypeptide being synthesised:

Original template strand:
C–**T**–C—C–T–T—T–T–T... → glutamic acid — glutamic acid — lysine...

Altered template strand: DNA:
C–**A**–C—C–T–T—T–T–T... → valine — glutamic acid — lysine...

The amino acid valine replaces the amino acid glutamic acid in the polypeptide. In fact, this is the mutation that results in the change in the β-chain of haemoglobin, which causes sickle-celled anaemia. The presence of a different amino acid in the polypeptide changes the structure of the haemoglobin — it is defective, and causes changes in the shape of red blood cells which, consequently, may clog blood vessels thus depriving vital organs of their full supply of oxygen.

Note that substitutions of the third base in the triplet codes may not affect the amino acid coded for because of the degenerate nature of the code (see Table 3).

In the following **base deletion**, guanine is lost from the template strand so that all the other bases move up by one. This is called a frameshift. The change affects not just one triplet code but all those thereafter. A multiple change in the amino acid sequence occurs and the final amino acid is missing:

Original template strand:
T–A–**G**—T–G–A—A–A–C—G–G–C... → isoleucine — threonine — leucine — proline

Altered template strand:
T–A–T—G–A–A—A–C–G—G–C...→ isoleucine — leucine — cysteine

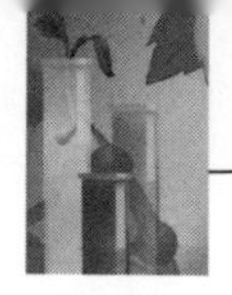

In the above example, the first triplet code is changed at the third base without affecting the amino acid it codes for. However, with a different amino acid sequence thereafter, the polypeptide produced would fold to form a different shape. If an enzyme was being produced it could lack the precise shape of the active site necessary for it to function. There are many examples of deletion mutations. For example, cystic fibrosis is caused by a deletion mutation, though in this case three bases are deleted so that one amino acid is missing. The effect is a non-functional membrane protein.

Gene mutations can be inherited, usually from carrier parents. However, a gene mutation in an organ giving rise to gametes (e.g. the ovary) will be passed on to offspring without any family history of the genetic disorder.

Mutations occur spontaneously and randomly. However, the rate of mutation can be increased by a number of factors including:

- carcinogenic chemicals — for example those in tobacco smoke
- high-energy radiation — for example ultraviolet radiation, X-rays

Mutation in a body cell (called a somatic mutation) may alter the way the cell functions and, indeed, may lead to its death. Alterations of certain genes may make the cell cancerous, i.e. the cell divides uncontrollably and produces a tumour.

Gene technology

Gene technology is the manipulation of genes for the benefit of people, in areas such as agriculture, food production and medicine. The tools and techniques of the genetic engineer are shown in Table 4.

Table 4 The tools and techniques of gene technology

Tool/technique	Purpose
Restriction enzymes	To cut DNA at specific points, making small fragments
DNA ligase	To join DNA fragments together
Vectors	To carry DNA into cells and ensure replication
Plasmids	Commonly used vectors
Genetic markers	To identify cells that have been transformed
PCR	To amplify very small samples of DNA
cDNA	To make a DNA copy of mRNA
DNA probes	To identify and label a piece of DNA containing a certain sequence
DNA sequencing	To read the base sequence of a length of DNA

Links Various DNA technologies, including restriction endonuclease enzymes, genetic markers, the polymerase chain reaction (PCR), DNA probes and DNA profiling (fingerprinting), are covered in the AS Unit 1 guide (pp. 38–43) in this series, to which you should refer.

Genetically modified organisms

Genetically modified organisms (GMOs) are also called genetically engineered organisms. Their production involves gene transfer, which is the transfer of a gene from a donor organism to a recipient organism.

Genetically engineered microorganisms (GEMs)

There are a number of stages in manipulating genes to produce genetically modified organisms and, specifically, genetically engineered microorganisms (GEMs):

- obtaining the required gene
- inserting the gene into a vector
- inserting the vector into a host cell
- identification of the host cells that have taken up the gene
- cloning the gene in the modified host cells

Obtaining the required gene

There are two main methods used to obtain a gene.

1 Restriction endonuclease enzymes are used to cut the gene out of chromosomal DNA (Figure 20):

- The chromosomal DNA is cut into fragments using an appropriate restriction endonuclease. The enzyme is 'appropriate' when it cuts either side of the gene.
- The DNA fragment containing the required gene is identified using a gene probe.
- A restriction endonuclease that cuts in a staggered fashion to produce 'sticky ends' is most useful. If the vector is opened with the same restriction enzyme, then the exposed bases of both are complementary and so will more readily attach through base pairing.

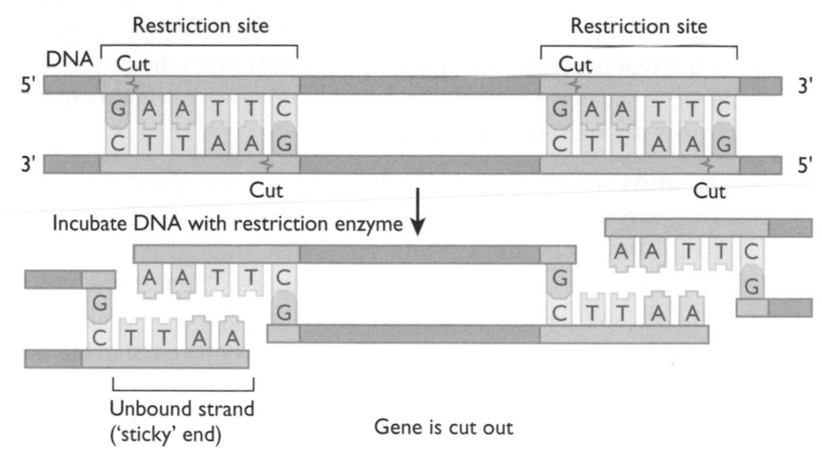

Figure 20 Use of restriction endonuclease enzymes to cut the gene out of chromosomal DNA

2 Reverse transcriptase is used to produce DNA from mRNA (Figure 21):

- Messenger RNA is obtained from cells where the gene concerned is actively synthesising protein (such cells possess many copies of the mRNA).

- The enzyme reverse transcriptase uses the mRNA as a template to produce a complementary single strand of DNA (complementary or cDNA) from free DNA nucleotides.
- Double-stranded DNA is made from the cDNA using the enzyme DNA polymerase.

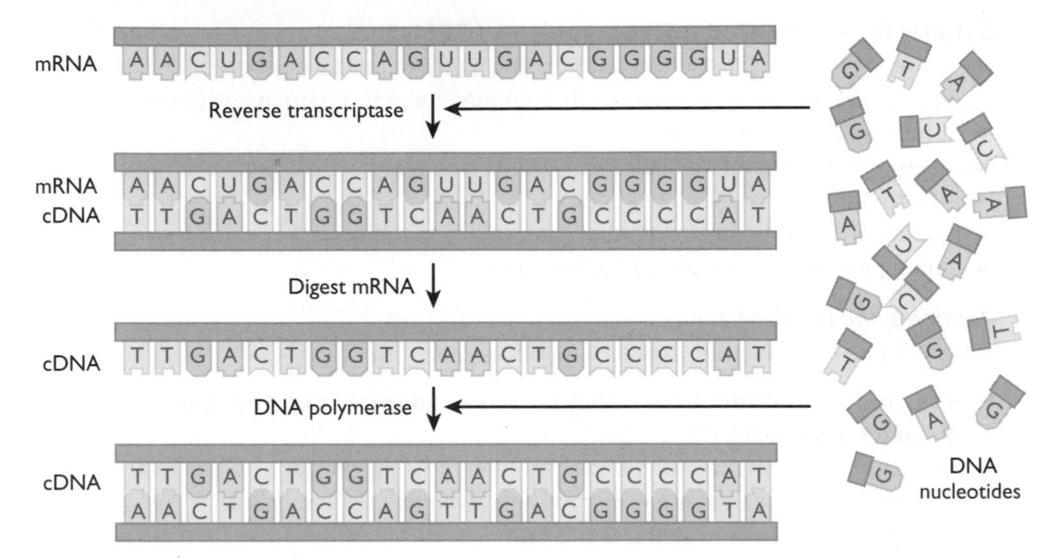

Figure 21 Use of reverse transcriptase to produce DNA from messenger RNA

Inserting the gene into a vector

Once the gene has been obtained it is inserted into a vector, which is a delivery tool to carry the gene into the host cell. A vector is needed because the length of DNA containing the gene will be ineffective since it is not part of the host cell's genome. In a vector, it can be replicated and expressed. The main types of vector are bacterial plasmids and viruses, though there are others.

1 Bacterial plasmids are the most common vector for inserting genes into bacterial cells. They are small circular pieces of DNA that occur in bacterial cells in addition to the main DNA. They may contain genes that are useful to the bacterium — for example genes that provide resistance to antibiotics. The method of inserting DNA containing the required gene into a plasmid involves the following processes:

- The plasmid is cut open using the same restriction enzyme used to cut the DNA fragment out of the donor DNA. The sticky ends of the two types of DNA contain complementary base sequences (if the gene was synthesised from mRNA, sticky ends are added).
- The plasmid DNA and gene DNA anneal (join). Hydrogen bonds form readily between the complementary bases of the sticky ends and **DNA ligase** catalyses the formation of covalent bonds (phosphodiester bonds) between the sugar-phosphate backbones of the plasmid DNA and the gene DNA. The gene is said to be **spliced** into the plasmid.

Any DNA that has 'foreign DNA' inserted into it is called **recombinant DNA**, so the plasmid is now a **recombinant plasmid**.

This process is shown in Figure 22.

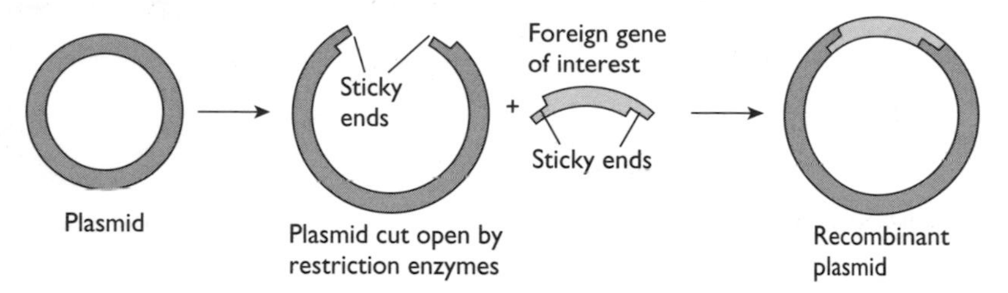

Figure 22 Transferring a gene into a plasmid

2 Viruses are adapted to insert their genetic material into a host cell — for example, a bacteriophage 'injects' its DNA into a bacterium. A bacteriophage that has a DNA fragment spliced into its DNA will transfer that recombinant DNA into the bacterial cell.

Inserting the vector into a host cell

Bacterial cells do take up plasmids — they are used naturally for the exchange of genes between bacterial cells. However, they do so more readily when induced. Cells are incubated with calcium ions and treated with a heat shock (temperature rise from 0°C to 40°C), which makes the cell wall permeable to plasmids. Bacteriophages are an effective way of delivering large genes into bacterial cells.

Identification of the host cells that have taken up the gene

Only a few of the bacterial cells will take up a recombinant plasmid — the rate of take-up may be as low as 1 in 10000. Most will either fail to take up a plasmid at all or will take up an original, non-transformed plasmid. Bacteria that have taken up the recombinant plasmids are called **transformed bacteria**. The two main ways of identifying such bacteria are marker genes and gene probes.

1 Using marker genes. Some plasmids carry genes that confer antibiotic resistance to the bacteria. The R-plasmid has genes for resistance to two antibiotics, ampicillin and tetracycline. These are used as marker genes. The restriction enzyme cuts in the middle of the tetracycline resistance gene and the 'desired' gene is inserted. The transformed plasmid now contains an active ampicillin resistance gene but an inactive tetracycline resistance gene (see Figure 23). Bacteria are detected according to the following criteria:

- Those that failed to take up plasmids are sensitive to both ampicillin and tetracycline.
- Those that take up the original plasmid are resistant to both ampicillin and tetracycline.
- Those that take up the recombinant plasmids are resistant to ampicillin but not to tetracycline.

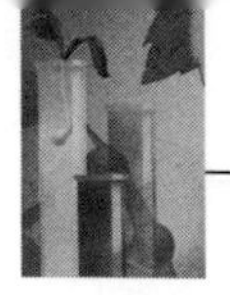

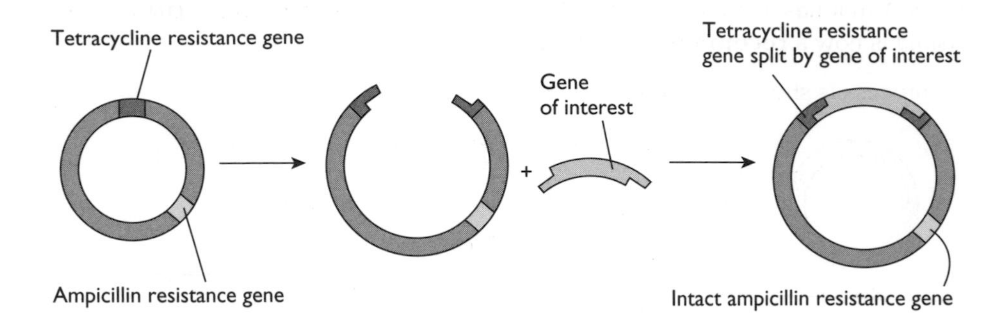

Figure 23 Inserting a gene into a plasmid with antibiotic resistant marker genes

The bacteria are cultured on agar plates. Each bacterium multiplies to form a colony. The colonies on the agar plate are then replica-plated onto a plate containing ampicillin and a plate containing tetracycline. Replica-plating involves 'blotting' the original plate with a pad and then pressing this against the surface of a fresh plate so that a few cells from each colony are transferred. The bacteria that survive on the ampicillin plate *only* are the transformed bacteria.

2 Using gene probes. A DNA or gene probe is a length of DNA — a 20-base sequence is sufficient — that is complementary to part of the gene being sought. Gene probes are produced so that each can base-pair with a complementary section of the DNA that makes up the gene. The gene probe and, therefore, the gene to which it can attach ('hybridise') can be identified if the probe is radioactively labelled. Figure 24 summarises the technique. Fluorescently labelled probes can also be used. They are detected using ultraviolet light.

Cloning the gene in the modified host cells

When the transformed bacteria are cultured their DNA replicates and the cells divide. As a result, many copies or **clones** of the desired gene are produced. These clones can be used to produce useful protein. This gene cloning technique can also be used to produce many copies of a healthy human gene, potentially for use in gene therapy.

Genetically modified plants

Foreign genes are inserted into plant cells using a range of methods.

- The most common method of transferring genes into plant cells is to use the common soil bacterium *Agrobacterium tumefaciens*, which readily invades damaged plant tissue and causes tumour-like growths. On entering the damaged tissue the bacterium's **tumour-inducing (Ti) plasmid** is transferred into a plant cell. For gene transfer, the desired gene is spliced into the plasmid, which is readily taken up by the plant cell provided that its cellulose wall has been removed by treatment with the enzyme cellulase.
- Minute pellets that are covered with DNA carrying the desired gene are shot through the cellulose walls into plant cells using a particle gun.

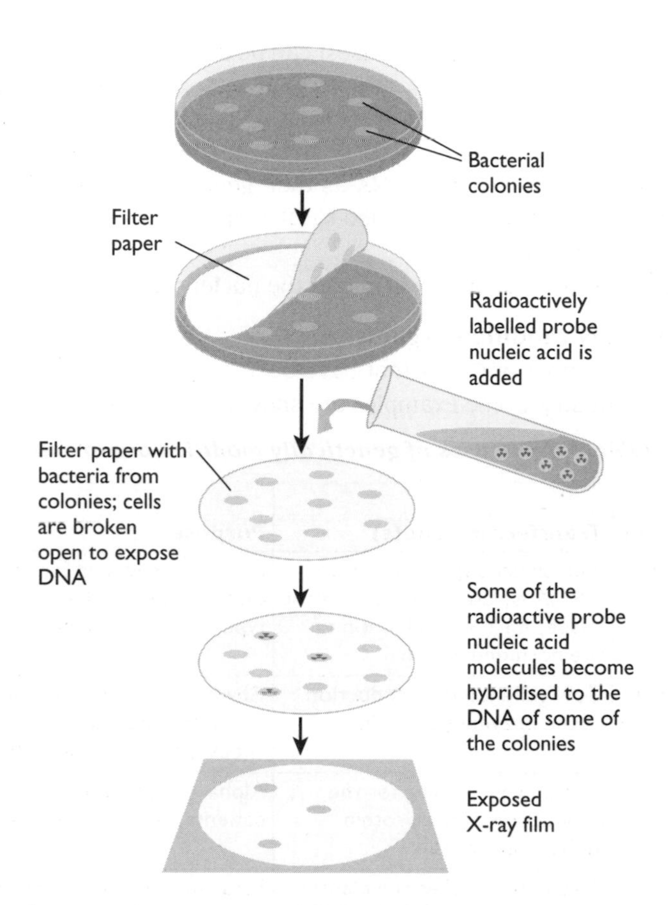

Figure 24 Using a radioactive gene probe to identify transformed bacteria with the desired gene

- Plant viruses are sometimes used. They infect cells by inserting their nucleic acid and, therefore, new genes can be transferred into the cell.

A gene for antibiotic resistance is inserted along with the desired gene. This is so that treatment with the antibiotic will kill the non-transformed cells, leaving only those that have been transformed. Transformed plant cells are then tissue-cultured and entire genetically modified plants are grown from this tissue.

Genetically modified animals

Several techniques can be used to artificially introduce genes into animal cells:

- A common method of introducing DNA into an animal cell is by **electroporation**, a technique in which the cell membrane is temporarily disrupted through treatment with a high-voltage shock.
- Another approach is to coat DNA in minute artificial lipid vesicles, called **liposomes**, which may adhere to the cell-surface membrane and pass the DNA into the cell in a similar way to endocytosis.

- **Viruses** (adenoviruses and retroviruses) can be used to insert genes into animal cells. Adenoviruses are human viruses that can cause respiratory diseases. Therefore, they have to be altered genetically so that the host cells are not destroyed. Adenoviruses are particularly useful for delivering genes to patients in gene therapy. Retroviruses have RNA as their genetic material. When their RNA is delivered into a host cell it is copied to DNA and the DNA is incorporated into the host's chromosome.
- It is possible to **inject DNA directly** into the nucleus of a fertilised egg.

Uses of genetically modified organisms

There are a huge number of GMOs that have been engineered for use in medicine, agriculture and industry. Some examples are shown in Table 5.

Table 5 Some uses of genetically modified organisms

Genetically modified organism	Transferred gene(s)	Purpose
Bacteria (*Escherichia coli*)	Human gene (produced from mRNA isolated from pancreatic tissue) for the production of the hormone insulin	Insulin is produced for use by people who cannot produce their own (suffer from type I diabetes mellitus)
Bacteria (*E. coli*) and fungi (*Aspergillus niger*)	Cow gene for the production of the enzyme chymosin	Chymosin is a coagulating enzyme used in the commercial production of hard cheese; most is genetically engineered
Yeast (*Saccharomyces cerevisiae*)	Human gene coding for the production of the protein alpha-1-antitrypsin	Alpha-1-antitrypsin is used to treat patients with hereditary emphysema
Soya (*Glycine max*)	A gene that makes the plants resistant to a specific herbicide (e.g. glyphosate)	Spraying with the herbicide kills competing weeds and so the productivity of the resistant crop plant is increased
Maize (*Zea mays*)	The *Bt* gene from the bacterium *Bacillus thuringiensis*	The gene produces a protein that is toxic to some insects (mainly caterpillars) so reducing the use of insecticides
Rice plants (*Oryza sativa*)	Two genes, each of which produces an enzyme involved in the metabolic production of beta-carotene (vitamin A)	Produces 'golden rice' with high levels of vitamin A used to supplement the diet in developing countries where people suffer permanent blindness from a lack of vitamin A
Cows	Human gene *LALBA* that codes for the synthesis of the protein alpha-lactalbumin	Produces human protein-enriched milk, more nutritionally balanced than natural cow's milk (for babies or the elderly with special nutritional needs)
Sheep (ewes)	Human gene for the blood-clotting protein Factor VIII	Factor VIII is extracted from the milk of transgenic ewes and used as a treatment for haemophilia
Chickens	Human gene for the production of the antibody mi-R24	The mi-R24 antibody is extracted from eggs and used in the treatment of the skin cancer malignant melanoma

Tip You do not have to learn a lot of examples of gene transfer but you should be familiar with some. Any example of a GMO in a question is likely to be unfamiliar. This is deliberate. You are being asked to apply your understanding of principles and procedures.

Gene therapy

Gene therapy is the treatment of a genetic disease by introducing the functional allele of the gene into the affected cells. Although attempts have been made to treat several different diseases (e.g. sickle-cell anaemia and muscular dystrophy) using gene therapy there are still major problems to be solved before treatments become sufficiently successful to be used widely.

Gene therapy for cystic fibrosis

Cystic fibrosis is a genetic disorder resulting from the mutation of a gene that codes for a carrier protein called CFTR. This protein lies in the cell-surface membrane of cells in many parts of the body, including the lungs, pancreas and reproductive organs. It transports chloride out of the cells. Water follows by osmosis. When the CFTR protein is not working, this does not happen. There is therefore much less water on the outer surface than there should be. The mucus that is produced in these areas therefore does not mix with water in the usual way. The mucus is thick and sticky. As a result:

- the abnormally thick mucus collects in the lungs, interfering with gas exchange and increasing the chance of bacterial infections
- the pancreatic duct may become blocked with sticky mucus, interfering with digestion in the small intestine
- reproductive passages (e.g. the oviducts) may become blocked, making a person sterile

Attempts have been made to treat cystic fibrosis by introducing the normal CFTR allele into the affected cells. Two methods of gene transfer have been trialled:

- inserting the normal allele into a modified adenovirus and then allowing the virus to infect cells in the person's respiratory passages — the virus enters the cells and so introduces the gene
- inserting the normal allele into liposomes and spraying these as an aerosol into the person's respiratory passages

In each case, there was limited success in that some of the cells lining the respiratory passages did take up the gene. However, there were also problems with these trials, including the following:

- Only a few cells took up the normal allele, so only those cells produced normal mucus.
- Cells in the surfaces of the respiratory passages do not live for long, so treatment would need to be repeated every few weeks.
- When using the adenovirus as a vector, some people developed serious lung infections.

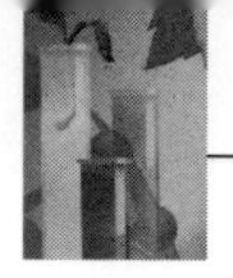

- It was only possible for cells in the respiratory passages to take up the normal allele, not cells in the pancreas or reproductive organs.

Gene therapy for SCID

Gene therapy has also been used to treat **severe combined immunodeficiency** (SCID), which is a rare autosomal recessive disorder. The first successful treatment of a child with SCID was in 1990. T-cells were removed, given the normal allele using a retrovirus as a vector, and the transformed T-cells were placed in her bone marrow. However, in later cases problems arose as patients developed a type of cancer called leukaemia.

Genome sequencing

The **genome** is the total genetic make-up of an organism, i.e. all of its DNA. The term originates from **gen**e and chromos**ome**. Genome is defined as the complete nucleotide (base) sequence in a haploid set of chromosomes of a eukaryotic cell. It is the complete set of genes, together with the non-coding DNA in between. Each genome contains the genetic coding needed to build and maintain that organism. **Genome sequencing** involves the laborious task of determining the order of nucleotides (bases) on each chromosome of a eukaryotic organism.

Genome sequencing projects have been undertaken for a range of organisms including the virus phage λ, the bacterium *Escherichia coli*, the plant *Arabidopsis thaliana*, the fruit fly *Drosphila melanogaster*, the mouse *Mus musculus* and humans. Those with smaller sequences were the first to be determined. The ultimate goal was the determination of the human genome sequence. Knowing the mouse genome is also important in that it is the model organism for much of the research into gene function.

Humans have 22 pairs of autosomes and two sex chromosomes. The **human genome project** worked on 24 separate chromosome sequences (22 autosomes and both X and Y chromosomes). The exact order of over 3 billion base pairs that make up these 24 chromosomes has been determined. Achieving this goal has helped to reveal an estimated 20000 to 25000 human genes in the total DNA. Many of these genes had been previously unknown. It has also revealed that less than 5% of the DNA represents the genes. There is a lot of non-coding DNA — approximately 95% of the genome.

Genes code for the synthesis of proteins. The non-coding DNA — most of it — has been referred to as 'junk' DNA. However, it seems likely that some of the non-coding regions (in and surrounding genes) contain signals that have not yet been recognised. Some DNA sequences might have other functions — suggestions include chromosomal replication, packaging the DNA into highly condensed chromatin and control of development. It seems that some of the non-coding DNA acts as genetic 'switches' that do not encode proteins but regulate when and where genes are expressed.

While a working draft of the human genome is available, researchers are now concentrating on adding the detail about human genes and the variations that can exist. The **HapMap project** aims to produce a map of human allele variants. Although any two unrelated people have about 99.9% of their DNA sequences in common, the remaining 0.1% is important because it contains the genetic variants that influence how people differ (e.g. in their risk of disease or their response to drugs). Sites in the genome where the DNA sequences of individuals differ by a single base are called single nucleotide polymorphisms (SNPs). About 10 million SNPs exist in human populations (where the SNP allele has a frequency of at least 1%).

Some of the *advantages of genome sequencing projects* are summarised below:

- A knowledge of the base sequence of genes allows the primary structure (amino acid sequence) of proteins to be determined. The use of molecular modelling software then allows the secondary, tertiary and quaternary structure to be predicted.
- A knowledge of the base sequence differences between different alleles (SNPs) should allow a better understanding of human diseases with a genetic cause. This in turn should facilitate their management and control through:
  - **gene therapy**, e.g. to treat cystic fibrosis
  - **genetic screening**. Genetic screening is used to determine if a person is at risk of passing on a hereditary disorder. The technique involves the use of 'DNA chips' — ordered series of nucleotide sequences that act as DNA probes. These allow the identification of a particular base sequence (specific to the 'disease' allele) and so determine if an individual is a carrier of a genetic disorder.
- Knowing the base sequences of different alleles (SNPs) provides improved diagnostics to test for the presence of genes that increase susceptibility of an individual to, for example, cancer or heart disease.
- Knowing the base sequences of different alleles (SNPs) provides an understanding of why different individuals respond differently to the same drugs. This may allow the development of 'designer drugs' matched to an individual's genetic profile.
- A number of genome sequencing projects are working on pathogenic micro-organisms. This should allow a better understanding of how microorganisms might act as pathogens. This should facilitate protection from the diseases they cause.
- An understanding of genome sequences should allow biologists to work out the various molecular interactions that lead to the normal development of organisms.
- Genome sequencing of different species allows direct comparisons between species. This will provide information that should allow evolutionary relationships to be determined.

Gene knockout technology

A **gene knockout** is a genetically engineered organism that carries a gene that has been made inoperative. This allows the function of genes to be studied. Comparing how the knockout organism differs from individuals in which that particular gene has not been made inoperative provides information about what the gene does. The

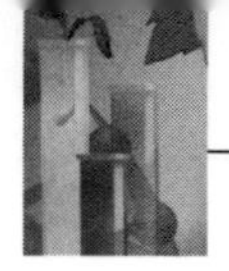

role of the gene in protein production and its metabolic and physiological influence can be determined. For example, imagine a gene GP, which makes the protein GP that controls the metabolism of glucose — a knockout model GP would be expected to lack the protein GP and have a problem with glucose metabolism.

Mice are the laboratory species most closely related to humans in which the knockout technique can be performed easily. A **knockout mouse** has both alleles of a particular gene inactivated. Knockout mice allow researchers to determine the role of a particular gene by observing what happens, both metabolically and physiologically, when gene function is lost. **Knockout models** (mice in which a specific gene has been 'knocked out') have been produced for studying human genetic diseases such as cystic fibrosis and thalassaemia — diseases in which the normal gene is defective. Mouse models have also been used extensively to study genes that become defective through mutation and cause cancer. More than 10 000 mouse genes (approximately half the genes in the mammalian genome) have been knocked out. The **knockout mouse project** aims to make knockout mice for all genes available for scientific and biomedical research.

Issues surrounding the use of gene technology

Some people find gene technology controversial. In such circumstances there is a need to be as fully informed as possible in order to evaluate and discuss the issues rationally. There are obvious potential *benefits* and, therefore, arguments in favour of its use. However, there are potential *risks* that may be cited in arguments against its use. Some examples are shown in Table 6.

In order to reduce risks from the use of genetically modified microorganisms a number of **safety precautions** have been devised:

- use of bacterial strains ill-adapted to the human physiology, for example:
  - strains that grow more slowly than normal wild-type intestinal bacteria so that they are out-competed and eliminated
  - strains with a minimum temperature tolerance above human body temperature so that they will not multiply in the human body
- use of strains that contain 'suicide genes' which are activated if conditions move outside certain pH or temperature limits
- use of containment mechanisms — for example, highly efficient air filters along with regular monitoring of the atmosphere in purpose-built laboratories

There are also **ethical concerns** over tampering with DNA of different species in ways that could never happen in nature. Some people are sceptical of the ownership of this powerful technology by a handful of multinational corporations that may be more interested in profits than in the long-term welfare of humans and the environment. There are also ethical concerns surrounding the use of genetic screening, i.e. the testing for genetic disorders in parents and embryos. Germ-line gene therapy — the transfer of genes into gametes to correct a genetic disorder — is a particularly contentious issue and raises the possibility of engineering 'designer babies'.

Table 6 Examples of the benefits and risks associated with the use gene technology

	Potential benefits: arguments for the use of gene technology	**Potential risks/concerns: arguments against the use of gene technology**
Genetically engineered micro-organisms (GEMs) developed to produce protein	More economic and wider production of medically important proteins, e.g. insulin	Some of the microorganisms (e.g. *E. coli*) live normally in the human gut; GEMs could escape from the laboratory and create a new stain of 'superbug'
Genetically modified (GM) plants including GM crops	Cheaper food for richer countries Possible reduction in the use of pesticides Reduction of food shortages in poorer countries	Risk of 'genetic pollution' with the spread of new genes from the modified crop to wild species, e.g. the formation of 'super weeds' Ecological concern that genetically modified plants may out-compete wild plants Concerns about allergic reactions
Genetically modified animals for food	Increased productivity of animals such as fish and cattle by transferring the gene for growth hormone into their genome	Concern that foreign protein, produced by transferred genes, may act as antigens (allergens) and increase the likelihood of allergies
Gene therapy	Effective treatment of genetic diseases (e.g. cystic fibrosis), relieving suffering and increasing life expectancy	Introducing genes into the human genome may disrupt the functioning of other genes, as in the appearance of leukaemia in patients treated for severe combined immunodeficiency (SCID)
Human genome research	Facilitates biomedical research	Concern that information from genome research might be used to produce 'designer babies' (e.g. for 'looks' and high IQ) Concern that an individual's genomic information (e.g. regarding susceptibility to heart disease) might become available to insurance companies
Genetic screening	People will have a better understanding of the risk of passing on a genetic disorder The fetus may be tested for the disorder before birth (called prenatal diagnostic testing)	Increased risk of stress resulting from the knowledge of being a carrier or of developing a disorder later in life (e.g. Huntington's disease) Termination of a pregnancy may not be acceptable
Gene knockout technology	Better understanding of how genes function; these genes might be implicated in a genetic disorder or might mutate to cause cancer.	Large numbers of mice are used in biomedical research, many of which may be in pain; there is the view that animals have rights and that it is unacceptable to use them in this way

With so many views, there is a need for the government to make decisions, i.e. there is a need for **legislation**. In the UK, the use of gene technologies is regulated strictly and research in the area of germ-line gene therapy in humans is banned.

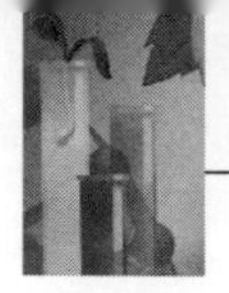

Tip You may expect to find questions that raise issues of an ethical nature. This is to emphasise that good scientific practice should consider not only what we 'can do' but also whether we 'should do' it. You are expected to appreciate and make informed comments on such aspects as: the ethical implications of the way research is carried out and the way in which society uses science to help in decision making. In any discussion on such issues you should be able to present a rational and balanced account, with arguments both for and against. You should find the website at **www.beep.ac.uk** helpful in developing your understanding.

Genes and patterns of inheritance

A **gene** is a length of DNA that codes for a particular trait. The position of a gene on the chromosome is called the **genetic locus** (plural: loci). For each gene there are alternative forms called **alleles**.

In most plants and animals, each body cell contains two sets of chromosomes. The chromosomes exist as homologous pairs. Homologous chromosomes have the same genetic loci, i.e. they possess alleles of the same genes (one from the mother and the other from the father). If the alleles on the homologous chromosomes are the same then the individual is **homozygous**; if the alleles are different the individual is **heterozygous**.

The two alleles for a particular trait represent the **genotype**. The actual appearance of the trait represents the **phenotype**. The phenotype is determined by the genotype but may also be modified by the alleles of other genes and by **environmental factors**.

Links The relationship between genes, chromosomes and ploidy is explained on pages 57 and 58 of the AS Unit 1 guide in this series, to which you should refer. In particular, you should study Figure 43: Genes, alleles, chromatids and homologous chromosomes.

How the phenotype is expressed depends on which allele is dominant. A **dominant** allele has its instruction followed and so its effect is produced in the heterozygous condition. The allele that does not have its instruction followed in the heterozygous condition is said to be **recessive**.

For example, humans possess a gene that determines the ability to taste phenylthiocarbamide (PTC). The PTC gene, TAS2R38, was discovered in 2003 as a consequence of work carried out during the Human Genome Project. There are two alleles: one (designated by the symbol **T**) is the tasting allele; while, the other (designated **t**) is the non-taster allele. The tasting allele (**T**) codes for a bitter taste receptor protein to which PTC can bind. The non-tasting allele (**t**) codes for a non-functional protein. **T** is dominant over **t**, since a heterozygote, **Tt**, possesses the allele **T** and produces the taste receptor protein. Things are never so simple, and environmental factors can affect PTC tasting ability — for example having a dry mouth may make it more difficult to taste PTC and what is eaten or drunk beforehand may also affect tasting ability.

Heredity of alleles at one locus: monohybrid inheritance

Heredity is the transfer of genetic factors from one generation to the next, i.e. from parents to their offspring. **Monohybrid inheritance** involves the inheritance of the alleles of a single gene.

In sexual reproduction, new individuals develop from a zygote produced by the fusion of male and female gametes. Since gametes are haploid (possess only one set of chromosomes) they only contain one allele of each gene. Thus, while a person has a pair of alleles for any genetic condition in body cells, only one allele of the pair is passed on via any one gamete. If an individual is homozygous (for example TT or tt in PTC tasting) then there can only be one type of gamete produced. If the individual is TT then all the gametes contain the allele T; if tt, then all the gametes contain the allele t. If an individual is heterozygous (for example Tt) then half the gametes produced will contain one allele (T) and half will contain the other (t). This is the essence of **Mendel's first law**, the law of segregation of factors, which states that '*when any individual produces gametes, the alleles separate, so that each gamete receives only one allele*'. This is explained by the separation of homologous chromosomes, carrying the alleles, during anaphase I of meiosis.

When gametes combine at fertilisation to form a zygote then the alleles are restored to a pair: one from the female parent (e.g. in an egg) and one from the male parent (e.g. in a sperm cell). Fertilisation is a random event. This means that any male gamete may fertilise a female gamete. The probability of alleles combining at fertilisation depends on the frequency of the gametes containing those alleles, For example, if half the male gametes contain **T** (because the male is heterozygous, **Tt**) and half the female gametes contain **T** (because the female is heterozygous, **Tt**) then the probability of a **TT** combination in a zygote is $\frac{1}{2} \times \frac{1}{2} = \frac{1}{4}$ (25%).

The analysis of patterns of inheritance, i.e. the way in which alleles are passed on from one generation to the next, depends on an understanding of these two phenomena:

- Alleles are separated in the production of gametes.
- Alleles combine at fertilisation when gametes fuse.

Different patterns of inheritance, with respect to the alleles of one gene, are provided by:

- dominance — the heterozygote has the same phenotype as the homozygous dominant genotype
- codominance — interaction of alleles results in a heterozygote with its own distinctive phenotype

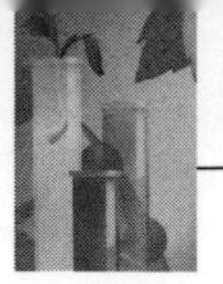

- lethal allelic combinations — one allele in the homozygous state causes death at an early stage
- multiple alleles — more than two alleles of a gene are possible
- sex linkage — the gene, and its alleles, are located on a sex chromosome, most often the X chromosome

Dominance

The Austrian biologist Gregor Mendel carried out breeding experiments in pea plants (*Pisum sativum*) and through these revealed the principles of inheritance. Mendel's work was successful because he bred plants through two generations and in large numbers so that the ratios achieved would be reliable. In one experiment, he crossed white-flowered and purple-flowered plants. He found that the $\mathbf{F_1}$ (first filial generation, or first generation of offspring) were all purple (we now know this is due to a dominant allele) but that when the F_1 were interbred the $\mathbf{F_2}$ (second filial generation) showed a mixture of purple and white-flowered plants in a ratio of 3:1.

In pea plants:

- the allele **P** codes for the production of purple pigment
- the allele **p** does not code for any pigment

The genotypes and phenotypes of flower colour in pea plants are shown in Table 7. The cross is explained in the genetic diagram in Figure 25.

Table 7 The genotypes and phenotypes of flower colour in pea plants

Genotypes	Phenotypes
PP	Purple flowers
Pp	Purple flowers
pp	White flowers

Parental phenotype	Purple flowers	White flowers	Plants from pure-breeding lines are cross-pollinated
Parental genotype	PP	pp	Both are homozygous
Parental gametes	(P)	(p)	Gametes are haploid, so contain only one allele from a pair — only one type of gamete from each parent
F_1 genotype	Pp		All F_1 plants are heterozygous with purple flowers — purple allele is dominant
F_1 gametes	(P) (p)	(P) (p)	All F_1 plants can produce two types of gamete — half with the purple flower allele and half with the white flower allele (two sets are shown to represent male and female gametes)

Fertilisations possible and F_2 genotypes:

	(P)	(p)
(P)	PP	Pp
(p)	Pp	pp

This is the standard way of showing all the possible fertilisations and possible combinations of alleles in the F_2 generation

F_2 phenotypes: 3 purple flowers : 1 white flower

In this instance, three of the possible four combinations contain at least one dominant allele and so have purple flowers, while only one of the four has two recessive alleles to give white flowers

Figure 25 A cross between pure-breeding white-flowered and pure-breeding purple-flowered pea plants

The test cross

Pea plants with purple flowers can be either homozygous (**PP**) or heterozygous (**Pp**). To find out which, a test cross is required. The individual to be tested (e.g. **PP** or **Pp**) is crossed with the homozygous recessive — in this case a plant with white flowers, genotype **pp**. Note that the homozygous recessive can be recognised by its phenotype and can only pass on a recessive allele, which will not influence the phenotype of the offspring. A test cross for purple flowering plants is shown in Figure 26.

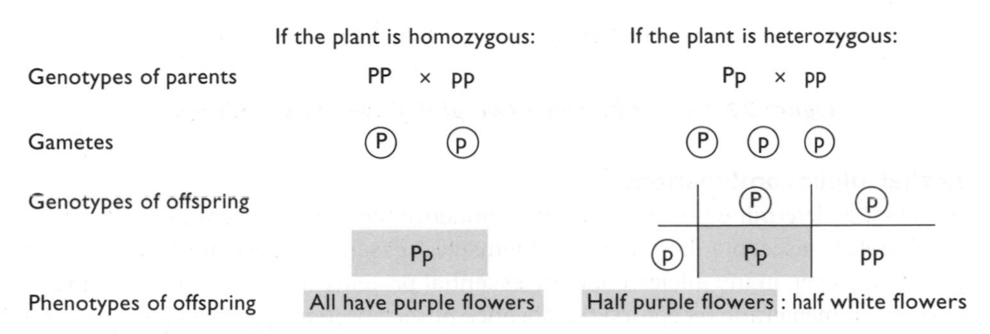

Figure 26 The test cross (testing the genotype of a purple-flowered plant)

If the individual tested is homozygous, all the offspring have the dominant phenotype. If the individual is heterozygous, the offspring will be a mixture of both dominant and recessive phenotypes.

Codominance (incomplete dominance)

In codominance a pair of alleles interact so that the heterozygote has a phenotype distinctive from (though often intermediate to) each homozygote. Codominance occurs because both alleles are functioning to produce something. An example is provided by pink flower colour in snapdragons (*Anthirrhinum*).

In snapdragons:

- the allele $\mathbf{C^R}$ codes for the production of red pigment
- the allele $\mathbf{C^W}$ codes for the production of white pigment

The genotypes and phenotypes of flower colour in snapdragons are shown in Table 8.

Table 8 The genotypes and phenotypes of flower colour in snapdragons

Genotypes	Phenotypes
C^RC^R	Red flowers
C^RC^W	Pink flowers
C^WC^W	White flowers

A cross between two heterozygotes produces a distinctive 1:2:1 ratio. This is shown in a cross between two pink-flowered snapdragons, explained in the genetic diagram in Figure 27.

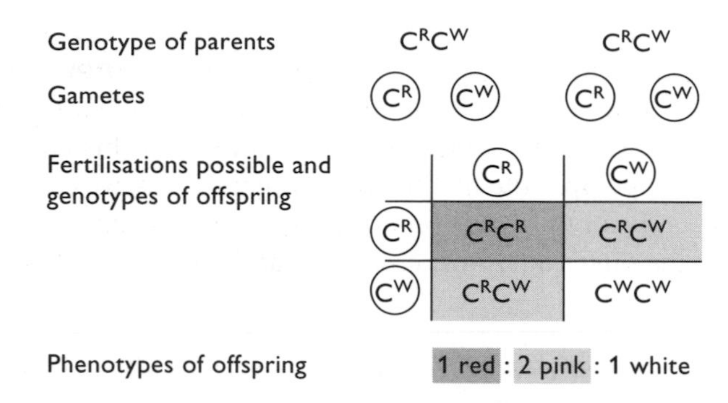

Figure 27 A cross between two pink-flowered snapdragons

Lethal allelic combinations

Alleles that determine flower colour may influence the attractiveness of the flower to a pollinating insect but they are not in themselves essential for the metabolism of the plant. However, many alleles code for essential proteins (e.g. enzymes, membrane carriers or membrane receptors) the absence of such proteins prevents the operation of vital functions. A deviant of such an allele is called a lethal allele. In general, the 'normal' allele has only to be present once (i.e. even in the heterozygous condition) for the protein to be produced and the vital function to be carried out. (Note that it is *normal* for alleles to differ and so the 'normal' allele is often called the wild-type allele in genetics.) Thus, most lethal alleles are recessive and so the lethal phenotype only occurs in individuals homozygous for the allele. Different types of lethality are recognised:

- The lethal allelic combination causes death of either the zygote or an early embryonic stage so there is no obvious evidence that the lethal allelic combination ever occurred.
- The lethal allelic combination causes death after a reduced lifespan. For example, Tay-Sachs disease is a rare genetic disorder of the central nervous system in humans. It is caused by the absence of the enzyme hexosaminidase (Hex-A). Without Hex-A, a lipid called GM2 ganglioside accumulates abnormally in cells, particularly in the nerves of the brain. The ongoing accumulation results in progressive damage to the nerve cells and death occurs in early childhood.
- The lethal allelic combination causes death at an early stage of development but its presence is evident in the heterozygote where it displays a distinctive phenotype. One example is provided by yellow coat colour in mice: the wild-type allele (**A**) codes for the production of a signalling protein involved in a range of functions, most noticeably the production of yellow and black bands on individual hairs (giving the agouti coat colour); it is also associated with proper embryonic development. The $\mathbf{A^Y}$ allele fails to code for this protein and is associated with obesity, diabetes and increased susceptibility to cancer, and a yellow coat colour from a failure to produce black pigment. The homozygous state, $\mathbf{A^Y\ A^Y}$, is lethal since it lacks the **A** allele for proper embryonic development.

In mice:

- the allele **A** codes for a signalling protein
- the allele **A^Y** does not code for this protein

The genotypes and phenotypes of coat colour in mice are shown in Table 9.

Table 9 The genotypes and phenotypes of coat colour in mice

Genotypes	Phenotypes
AA	Agouti coat colour
AA^Y	Yellow coat colour
A^YA^Y	Lethal: embryonic death

A cross between two heterozygotes produces a distinctive 2:1 ratio. This is shown in a cross between yellow mice, explained in the genetic diagram in Figure 28.

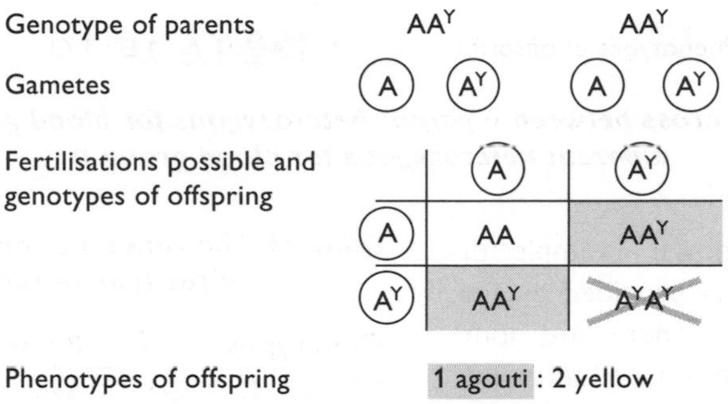

Genotype of parents: AA^Y × AA^Y

Gametes: (A) (A^Y) (A) (A^Y)

Fertilisations possible and genotypes of offspring:

	A	A^Y
A	AA	AA^Y
A^Y	AA^Y	~~A^YA^Y~~

Phenotypes of offspring: 1 agouti : 2 yellow

Figure 28 A cross between yellow mice

Multiple alleles

Multiple alleles describes the situation where more than two alleles of a particular gene exist in the population (of course only two can exist in any one individual). An example is the inheritance of the ABO blood group which also demonstrates codominance and dominance: two alleles are active in coding for the production of the A-antigen and the B-antigen and so are codominant; a third allele does not code for any antigen and so is recessive to the other two alleles. The genotypes and phenotypes are shown in Table 10.

In blood group inheritance in humans:

- the allele $\mathbf{I^A}$ codes for the production of antigen A
- the allele $\mathbf{I^B}$ codes for the production of antigen B
- the allele **i** does not code for any antigen

Table 10 The genotypes and phenotypes of blood groups in humans

Genotypes	Phenotypes
I^AI^A, I^Ai	Blood group A
I^BI^B, I^Bi	Blood group B
$I^A I^B$	Blood group AB
ii	Blood group O

A cross between a parent heterozygous for blood group A and a parent heterozygous for blood group B has the potential to produce children with any of the four blood groups. This is shown in Figure 29.

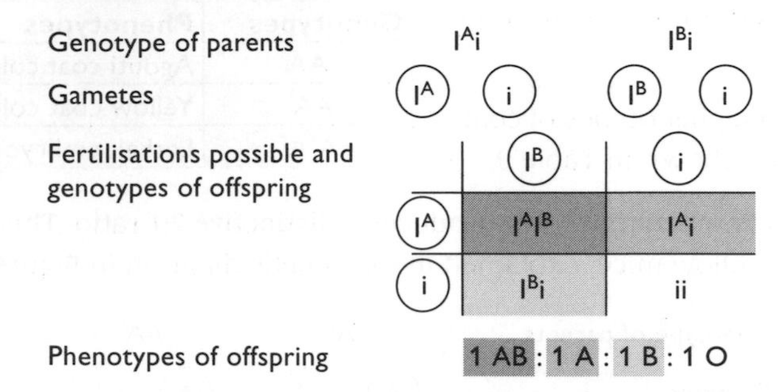

***Figure 29** A cross between a parent heterozygous for blood group A and a parent heterozygous for blood group B*

Another well-known example of multiple alleles is provided by fur type in rabbits. There are four alleles with a dominance hierarchy:

- **C** — dominant to $\mathbf{c^h}$, $\mathbf{c^{ch}}$ and $\mathbf{c^a}$
- $\mathbf{c^h}$ — dominant to $\mathbf{c^{ch}}$ and $\mathbf{c^a}$
- $\mathbf{c^{ch}}$ — dominant to $\mathbf{c^a}$
- $\mathbf{c^a}$ — recessive

***Table 11** The genotypes and phenotypes of fur type in rabbits*

Genotypes	Phenotypes
CC, Cc^h, Cc^{ch}, Cc^a	Agouti fur
c^hc^h, c^hc^{ch}, c^hc^a	Himalayan fur
$c^{ch}c^{ch}$, $c^{ch}c^a$	Chinchilla fur
c^ac^a	Albino fur

The genotypes and phenotypes for fur type in rabbits are shown in Table 11.

Only a rabbit with albino fur has a known genotype — $\mathbf{c^ac^a}$. Others would require a test cross to ascertain their genotype, i.e. cross the individual with the bottom recessive, $\mathbf{c^ac^a}$ (an albino rabbit).

Sex linkage

In mammals, one pair of chromosomes is associated with gender. These are the sex chromosomes, designated **XX** in females and **XY** in males. (The other, non-sex chromosomes are called autosomes.) All eggs contain an X chromosome, half of the sperm contain an X chromosome and half contain a Y chromosome.

The X and Y chromosomes are largely non-homologous and so carry different genes. Indeed, the Y chromosome carries very few genes and these are, in general, concerned with the development of maleness — for example, the gene SRY triggers testis development. Sex-linked inheritance occurs with the alleles of genes located on the sex chromosomes. **Y-linked inheritance** is confined to males and is relatively rare. **X-linked traits** are relatively common.

Fruit flies (*Drosophila melanogaster*) also have XX females and XY males. An example of an X-linked recessive trait is 'white eye'. Genotypes of sex-linked traits include the appropriate chromosome, X or Y, as well as the allele. The dominant wild-type allele,

X^+, determines red eye, while X^W represents the recessive white eye allele. There is no equivalent allele on the Y chromosome.

In fruit flies:

- the wild-type allele X^+ codes for red eye and is dominant
- the allele X^W codes for white eye and is recessive

Table 12 The genotypes and phenotypes of white eye in fruit flies

Genotypes	Phenotypes
X^+X^+, X^+Y	Red-eyed female and red-eyed male
X^+X^W	Red-eyed female (carrier)
X^WX^W, X^WY	White-eyed female and white-eyed male

The genotypes and phenotypes for eye colour in fruit flies are shown in Table 12.

Figure 30 shows reciprocal crosses between red-eyed and white-eyed flies:

- a red-eyed female and a white-eyed male
- a white-eyed female and a red-eyed male

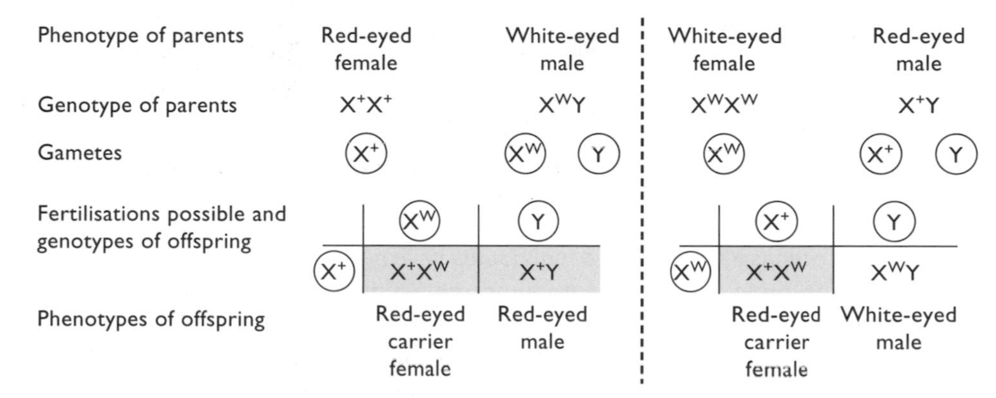

Figure 30 Reciprocal crosses between red-eyed and white-eyed fruit flies

X-linked recessive traits share the following features:

- Reciprocal crosses produce different results and are used in testing for sex linkage.
- They are more common in males than females. This is because the allele only has to appear once in the male but has to be inherited from both parents in an affected female.
- Affected males inherit the allele from the female parent since the Y chromosome is inherited from the male parent (and most commonly from a heterozygous — carrier — female parent).
- Affected females inherit the allele from both parents (and so the male parent must also be affected).

There are a number of X-linked recessive traits in humans — for example, red–green colour blindness and haemophilia. In cats there is the interesting example of a coat colour called tortoiseshell. The condition is codominant and so tortoiseshell cats are heterozygotes. It only occurs in females since males cannot have both alleles.

X-linked dominant traits exist, such as Huntington's chorea in humans. The trait is passed from an affected father to all his daughters (since a daughter always inherits her father's X chromosome).

The XX/XY sex-determination system is not universal. For example, in birds, the female is XY and the male is XX, and in crocodiles it is the temperature at which eggs develop that determines gender.

Human genetics and pedigrees

A pedigree charts the transmission of a genetic trait over several generations in a family. Pedigrees can be used to analyse the pattern of inheritance of genetic disorders. Two pedigrees are illustrated in Figure 31.

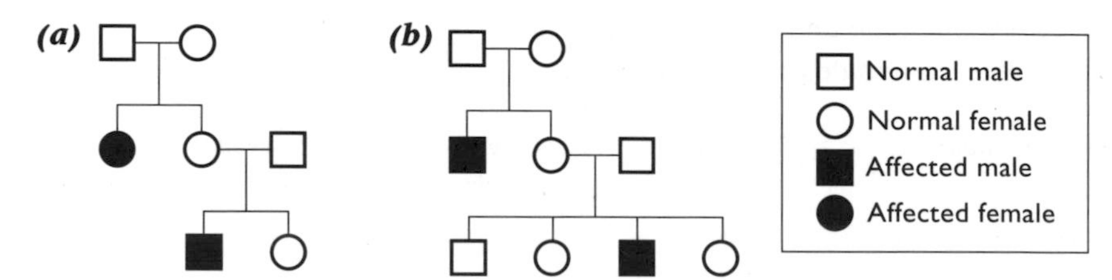

Figure 31 Pedigrees for (a) cystic fibrosis, and (b) Duchenne muscular dystrophy

Both traits are recessive — affected individuals have parents without the trait. Cystic fibrosis is autosomal recessive — it cannot be X-linked since an affected female has a father who is unaffected (see Figure 31(a)). Duchenne muscular dystrophy is X-linked recessive — Figure 31(b) shows that the trait is more common in boys. However, a more extensive pedigree would be required to determine X-linkage with any degree of certainty.

Considerations when analysing pedigrees include:

- **autosomal recessive** — parents of an affected individual may not be affected, i.e. the trait may appear to skip a generation
- **X-linked recessive** — as for autosomal recessive but:
  - more common in males than females (passed on from carrier mothers)
  - for a female to be affected the father must also be affected
  - an affected female will pass the trait on to all her sons
- **autosomal dominant** — at least one parent must be affected (though if both parents are affected not all the children would necessarily be affected, since the parents could be heterozygous)
- **X-linked dominant** — as for autosomal dominant but an affected father will pass the trait on to all his daughters

Heredity of alleles at two loci: dihybrid inheritance

Mendel also performed crosses in which he followed the segregation of the alleles of two genes. These experiments formed the basis of **Mendel's second law**, the law of independent assortment, which states that *'during gamete formation, the segregation of the alleles of one gene is independent of the segregation of the alleles of another gene'*. This is explained by the random arrangement of the homologous pairs on the equator of the spindle at metaphase I of meiosis and their subsequent separation during anaphase I.

Links The meiotic events that explain Mendel's laws of inheritance are covered on pages 60–61 of the AS Unit 1 guide in this series, to which you should refer.

Dominance at each of the two loci

Figure 32 shows the results of a cross between a pure-breeding tall, purple-flowered pea plant with a pure-breeding short, white-flowered pea plant through to the F_2 generation.

Parental phenotype	Tall, purple	Short, white	Plants from pure-breeding lines are cross-pollinated
Parental genotype	TTPP	ttpp	Both are homozygous
Parental gametes	(TP)	(tp)	Gametes are haploid, so contain only one allele from a pair — only one type of gamete from each parent
F_1 genotype	TtPp		All F_1 plants are heterozygous tall, purple — tall and purple alleles are dominant
F_1 gametes	(TP) (Tp) (tP) (tp)		All F_1 plants can produce four types of gamete — either of the 'height' alleles may be separated with either of the 'flower colour' alleles (only one set is shown)

Fertilisations possible and F_2 genotypes

	(TP)	(Tp)	(tP)	(tp)
(TP)	TTPP	TTPp	TtPP	TtPp
(Tp)	TTPp	TTpp	TtPp	Ttpp
(tP)	TtPP	TtPp	ttPP	ttPp
(tp)	TtPp	Ttpp	ttPp	ttpp

With four types of gamete, there are 16 possible combinations — nine have both dominant traits, three are tall but white, three are short but purple, while only one of the 16 has both recessive traits

F_2 phenotypes 9 tall purple : 3 tall white : 3 short purple : 1 short white

Figure 32 A cross between pure-breeding tall, purple-flowered and pure-breeding short, white-flowered pea plants

Independent assortment only works for genes located on different chromosomes. However, chromosomes, particularly those that are longer, are composed of many genes — a genetic situation called linkage.

Tip You do not need to know about the inheritance of linked genes.

Different genetic situations

In the situation described above, there were dominant alleles at each genetic locus. However, codominance, lethal allelic combinations, multiple alleles and sex linkage are all possible — for example, there may be codominance at one genetic locus and sex linkage at another.

Additive effect of the alleles at two loci

Many traits are governed by the cumulative affects of two or more genes. For example, grain colour in wheat (*Tritium aestivum*) is determined by the additive effects of the alleles of two genes, **A/a** and **B/b**. The more **A** and **B** alleles that are present in a genotype the deeper the shade of red in the grain. The absence of **A** and **B** alleles causes the grain to be white. Indeed, five shades of colour are possible: darkest-red, dark-red, red, light-red and white. The results of a cross between two individuals of genotype AaBb, red grain colour, are shown in Figure 33.

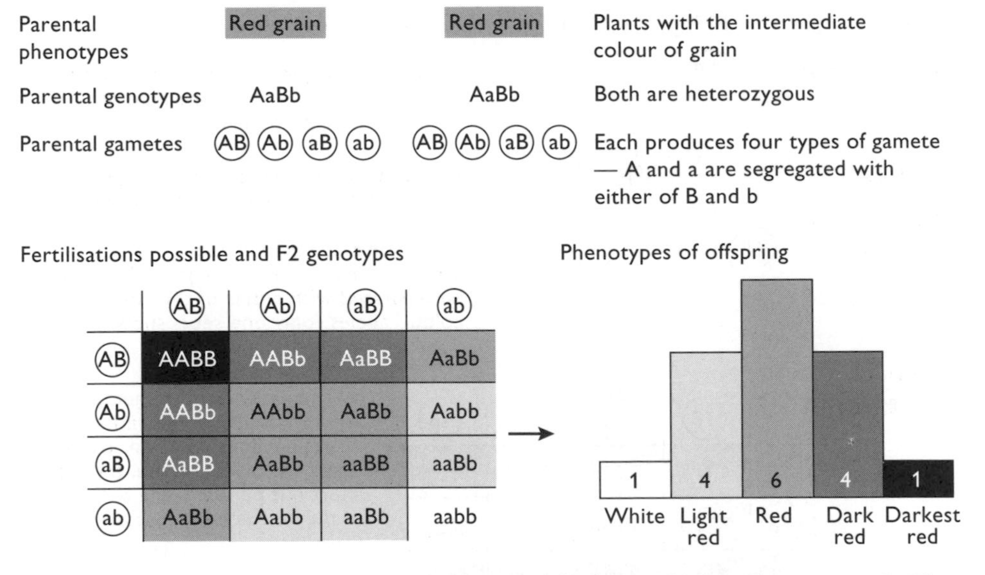

Figure 33 A cross between two red-grained wheat plants of genotype AaBb

Where more than two genes are involved, polygenic inheritance results. This explains quantitative characteristics (such as height or weight) that exhibit the normal frequency distribution for a continuous variable, particularly when the further influence of the environment is added.

Gene interaction: epistasis

Epistasis is the situation where one gene interferes with the expression of another. For example, in squash plants fruit colour is determined by the interaction of two genes, **W/w** and **G/g**. The **W** allele suppresses the action of the **G/g** gene so the fruit are white. When the genotype is **ww**, the **G/g** gene is expressed, with the dominant **G** allele producing yellow fruit colour. The **g** allele is expressed recessively as green.

A cross between two squash plants of genotype **WwGg** produces an offspring genotypic ratio of 9 **W–G–** : 3 **W–gg** : 3 **wwG–** : 1 **wwgg**, and so the phenotypic ratio is 12 white : 3 yellow : 1 green.

There are many examples of epistasis. Indeed, purple flower in pea plants involves the action of two enzymes each produced by a different gene:

		gene A		gene B	
Precursor (colourless)	→		Intermediate (colourless)	→	Purple pigment

A cross between two purple-flowered plants of genotype **AaBb** would produce an offspring genotypic ratio of 9 **A–B–** : 3 **A–bb** : 3 **aaB–** : 1 **aabb**, and so the phenotypic ratio is 9 purple : 7 white.

A further example is provided by coat colour in mice. The agouti (banded hair) allele, **A**, is dominant to the black allele, **a**. The coat colour alleles are only expressed if the **E** allele of an epistatic gene is present, i.e. they are not expressed if the genotype is **ee** and a white coat is produced. A cross between two agouti mice of genotype **AaEe** would produce an offspring genotypic ratio of 9 **A–E–** : 3 **aaE–** : 3 **A–ee** : 1 **aaee**, and so the phenotypic ratio is 9 agouti : 3 black : 4 white.

Tip Here are some tips for doing genetics problems. They are generally of two types:

(1) Information regarding the genetics of the situation (e.g. alleles involved, dominance) is presented and the question involves working out the offspring genotypes and phenotypes resulting from a particular cross:

- Present the symbols representing the alleles (if this has not already been done) — for example, **A** = one allele; **a** = alternative allele.
- Present the genotypes and their corresponding phenotypes — for example, **Aa** = what phenotype? Here, this depends on dominance and may be part of the question.
- Present the cross thus (and as provided in previous figures):
  - **P** Show parental genotypes (and phenotypes).
  - **Gametes** Show the *types* of gametes produced, i.e. apply the laws of inheritance. Only one allele of the pair enters a gamete. If homozygous then there is only one type of gamete; if heterozygous two types are produced. Either allele of one gene may be separated with either allele of another gene. If homozygous at both genes then there is one type of gamete; if homozygous at one gene and heterozygous at the other two types are produced; if heterozygous at both loci then four types are produced.
  - **Fertilisations** — use a Punnet square if there are two (or more) types of gamete from each parent.
  - **Offspring** — summarise the offspring genotypes and phenotypes and their proportions.

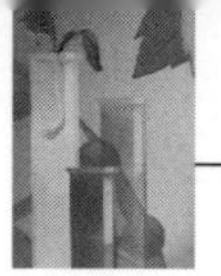

(2) Information on a particular cross (e.g. the phenotypes of the parents and offspring) is presented and the question involves working out the genetics of the situation — for example, which allele is dominant, is there evidence of lethality or sex linkage?

- Summarise the information given in a flow diagram. For example:

  taster × taster → tasters and non-tasters

- Note that pure-breeding individuals are *homozygous* for the particular character and something that does not breed true is *heterozygous* — for example, if both parents are tasters and have a non-taster son or daughter then the parents are heterozygous.
- A phenotype that exhibits the recessive trait is homozygous — for example, the dominant trait (tall) is **AA** *or* **Aa**, while the recessive trait (short) must be aa.

Note that the phenotypic ratios can be distinctive:

- With the alleles of *one* gene — a 3:1 ratio indicates two types of gamete from each parent both of which must be heterozygous; 1:1 indicates two types of gamete from one parent, which must be heterozygous and one type from the other, which must be homozygous; 1:2:1 indicates codominance with both parents heterozygous; 2:1 is indicative of a lethal combination where both parents are heterozygous.
- With the alleles of *two* genes — a 9:3:3:1 results from a 4 × 4 Punnet square, so both parents must be heterozygous at each locus and produce four types of gamete; this is also the case where any ratio 'adds' up to 16 (such as 9:7, 9:3:4 or other 'odd' ratios indicating gene interaction); 3:1:3:1 results from a 2 × 4 Punnet square, so one parent is heterozygous at one locus and homozygous recessive at the other and the other parent is heterozygous at both; 1:1:1:1 may result from a 2 × 2 Punnet square where one parent is heterozygous at one locus and homozygous recessive at the other, while the other parent has the reverse situation (e.g. **Aabb** × **aaBb**) *or* may result from a 1 × 4 Punnet square where one parent is homozygous recessive at both loci and the other parent is heterozygous at both loci (e.g. **aabb** × **AaBb**).

Population genetics, evolution and speciation

Population genetics

A population is a group of individuals of the same species living in the same habitat because they are similarly adapted. Members of the population are capable of interbreeding and so their genes and alleles are free to mix. All the genes and alleles in a population at a particular time make up the **gene pool**. The proportion of a particular allele in the population is called the **allele frequency**.

Consider a single gene that has two alleles — **A** and **a** — in the population. The frequencies of the alleles, **A** and **a**, are represented by the algebraic symbols p and q, respectively. Since **A** and **a** are the only alleles of that gene, their combined frequencies must equal 1:

$$p + q = 1$$

Genotypes are formed from the fusion of gametes, each of which contains a single allele, **A** or **a**. If the probability of the different alleles combining at fertilisation depends solely on their respective frequencies in the gene pool, then the **genotype frequencies** are determined as:

genotype **AA** has frequency of p^2

genotype **aa** has a frequency of q^2

genotype **Aa** has a frequency of $2pq$

The condition for allowing this determination is often described as **random mating** (with respect to the alleles in question). Again, since these are all the genotypes in the population, their sum must equal 1:

$$p^2 + 2pq + q^2 = 1$$

This is called the **Hardy–Weinberg equation**.

The Hardy–Weinberg equation relates allele frequency to genotype frequency, i.e. it allows for the determination of one from the other. If the allele frequencies (p and q) are known, then the genotype frequencies can be determined. It is, however, more likely that the phenotype frequencies are known (since these are what can be seen). Even then, the equations can be used to determine the frequencies of the alleles and of the different genotypes. For example, the cystic fibrosis (CF) gene has two alleles, the normal F allele that leads to normal mucus production and the recessive **f** allele that leads to the production of thicker mucus and, hence, cystic fibrosis. The pair of alleles of the CF gene has three possible combinations — homozygous dominant (**FF**), homozygous recessive (**ff**) and heterozygous (**Ff**). Both **FF** and **Ff** (carriers) have the normal phenotype; **ff** has cystic fibrosis. In Ireland, the incidence of cystic fibrosis is around 1100 in a total population of 5.5 million. This represents a frequency of 0.0002 (or 0.02%, though it is better to express frequencies as decimal fractions because the arithmetic is more straightforward). If the frequency of the homozygous recessive, **ff**, is equal to q^2 then:

$$q = \sqrt{0.0002} = 0.014$$

$p = 0.986$ (since $p + q = 1$)

frequency of **FF** $= p^2 = 0.972$

frequency of **Ff** $= 2pq = 0.028$

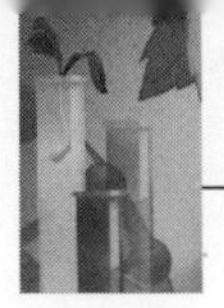

So the frequency of those who carry the **f** allele in Ireland is 0.028. This might seem small but it means that 0.028 multiplied by 5.5 million, i.e. 154 000, is the number who carry the **f** allele. Viewed locally, it means that in a school population of 1000, 28 students might be *expected* to be carriers.

Tip These are suggested steps in calculating allele and genotype frequencies in population genetics problems, dependent on the information initially available.

Knowing the allele frequencies

The frequencies of the alleles may be presented. Determining the frequencies of the genotypes and phenotypes should then be relatively straightforward. The steps in solving problems are:

- Present the frequencies of both alleles (p and q) remembering that $p + q = 1$.
- Calculate the frequency of the homozygous dominants as p^2.
- Calculate the frequency of the heterozygotes as $2pq$.
- Calculate the frequency of the homozygous recessives as q^2. This is also the frequency of the recessive trait. The actual number of homozygous recessive individuals in the population can be calculated as $q^2 \times N$ (where N is the total number of individuals in the population).
- Calculate the frequency of the dominant trait as $p^2 + 2pq$. The actual number of individuals with the dominant trait in the population can be calculated as $(p^2 + 2pq) \times N$ (where N is the total number of individuals in the population).
- The actual number of homozygous dominant individuals in the population can be calculated as $p^2 \times N$ (where N is the total number of individuals in the population).
- The actual number of heterozygous individuals in the population can be calculated as $2pq \times N$ (where N is the total number of individuals in the population).

Knowing the frequencies of phenotypes

In general, phenotype frequencies can be determined, since it is the phenotypes that can be recognised in a population. It must be remembered, however, that the dominant trait consists of both homozygotes and heterozygotes. Only the homozygous recessive individuals contribute towards the recessive trait.

frequency of dominant trait = $p^2 + 2pq$

frequency of recessive trait = q^2

It is difficult to work arithmetically with percentages, so frequencies must always be presented as proportions, i.e. as decimal fractions. The steps in solving problems are:

- Determine the frequency of the recessive trait. Ensure that it is presented as a proportion (decimal fraction). This is q^2.
- Calculate q (as the square root of q^2). This is the frequency of the recessive allele (e.g. **a**).
- Calculate p (as $p = 1 - q$). This is the frequency of the dominant allele (e.g. **A**).
- Calculate p^2. This is the frequency of the homozygous dominant individuals in the population. The actual number of homozygous dominant individuals in the

population can be calculated as $p^2 \times N$ (where N is the total number of individuals in the population).

- Calculate $2pq$. This is the frequency of the heterzygotes in the population. The actual number of heterozygous individuals in the population can be calculated as $2pq \times N$ (where N is the total number of individuals in the population).

While the Hardy–Weinberg equation is useful for studying genes in populations, it must be remembered that its use entails certain assumptions. The Hardy–Weinberg principle is that no factor operates that would cause the allele frequencies in the offspring generation to differ from those in the parental population. This means that allele frequencies remain constant over time — the population is said to be in Hardy–Weinberg equilibrium. The conditions for this to happen include:

- **Mating is random**. In fact, non-random mating causes changes in the proportions of genotypes without affecting allele frequencies — for example, inbreeding promotes homozygosity (more alleles than expected appear in the homozygous state).
- **The population is large**. In small populations random changes can disrupt the allele frequencies, a situation called genetic drift.
- **No mutations are occurring**. In general, the rate of gene mutation is so low (in the order of 1 in 10 million) that it has little impact on allele frequencies. (The significance of mutation is the introduction of new alleles into populations.)
- **No migration is taking place** (either into or out of the population).
- **No selection is taking place** (i.e. all alleles are equally advantageous). Selection is the principle agent influencing changes in the frequency of alleles in populations.

If it is found that the allele frequencies of a gene pool are changing over time, then it is evidence that selection is operating in the population.

Selection

Selection and genetic variation

Populations are genetically variable. There are a number of reasons for this:

- **Mutation** is a source of genetic variation. **Gene mutation** introduces new alleles into the population. **Chromosome mutation** occurs when chromosomes fail to separate during meiosis so that extra (or occasionally fewer) chromosomes appear in the affected individual. There are two types:
  - **aneuploidy**, in which one chromosome only fails to separate and the affected individual has either an extra chromosome (as in Down's syndrome with an extra chromosome 21) or a missing chromosome (Turner's syndrome, XO, with only one X chromosome)
  - **polyploidy**, in which the whole set of chromosomes fails to separate and the affected individual has three (or more) sets of chromosomes rather than the normal two sets (diploid) — for example, the Bramley apple has three sets of chromosome (triploid)

- **Meiosis** (whether in the production of gametes in animals, or in the production of spores in plants) produces a variety of haploid cells through independent assortment and recombination.
- **Sexual reproduction**, particularly when cross fertilisation is ensured, is the most important factor promoting genetic variability in populations. It involves the mixing of genetic material from the haploid cells of two different individuals. In life cycles, meiosis (production of haploid cells) and sexual reproduction act to shuffle the allelic combinations present previously.
- Variability is preserved by diploidy which shelters rare, recessive alleles, i.e. *heterozygotes act as important reservoirs of genetic variation* in populations.

In a genetically variable population, some individuals will possess certain alleles, or combinations of alleles, that make them better adapted to a particular environment. These individuals survive and reproduce more successfully than others in that environment. This is **natural selection**, which is defined as *'the unequal transmission of alleles to subsequent generations by different genotypes'*.

Polymorphic populations may be used to study natural selection. Polymorphism is the presence in a population of two or more distinct forms. For example, the snail *Cepaea nemoralis*, possesses shells which vary in colour (brown, pink or yellow) and in the number of bands (five, three or none). Studies have shown that thrush predation is a powerful agent of selection: in grassland, yellow, banded forms are selected for as they are better camouflaged; on the floors of beech forest, brown, unbanded snails are favoured. The result is considered to be a balanced polymorphism. Different forms can also be maintained in a population by a type of selection called apostatic selection — a predator hunts the most common form (since it is easier to see) until it becomes the less common form. At this point, another form becomes the favoured prey. Heterozygous advantage is sometimes suggested as a means of maintaining different forms in a population. For example, if **Aa** was selected for (and both **AA** and **aa** disadvantaged) then, while **Aa** would increase in frequency, it would always interbreed to produce more homozygotes.

Natural selection can act to remove some of the variants from a population and so reduce the amount of genetic variation. There are different ways in which selection might do this, including stabilising selection and directional selection.

Links You should revise the descriptions of stabilising and directional selection provided in the AS Unit 2 guide, pp. 58–59, in this series.

Stabilising selection

Stabilising selection occurs where *environmental conditions are largely unchanging*. It favours the modal or intermediate forms and acts against the extremes.

An example of stabilising selection comes from human birth records of babies born in London between 1935 and 1946 (see Figure 34). It shows that there is an optimum birth weight for babies, and that babies with birth weights heavier or lighter are at a selective disadvantage.

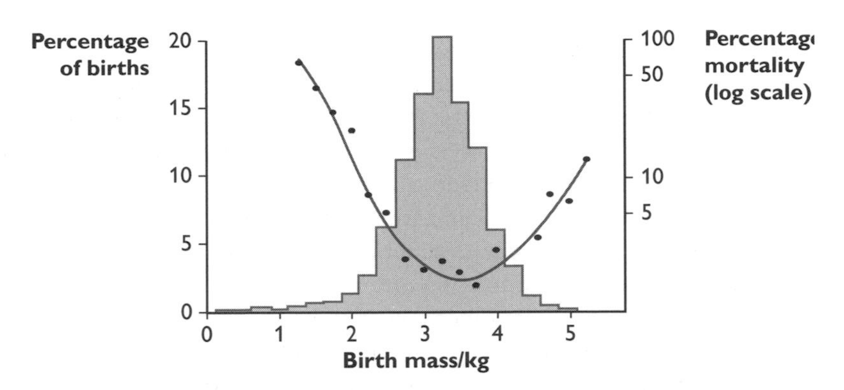

Figure 34 Stabilising selection in size of human babies

Stabilising selection does not lead to evolution, it maintains the **adaptive norm**.

Directional selection and evolutionary change

Directional selection is associated with *changing environmental conditions*. In these situations, the majority of an existing form may no longer be best suited to the environment. Some extreme forms of the population may have a selective advantage (i.e. be more likely to survive and breed) in the changed conditions. They will therefore contribute more offspring, and the alleles these offspring possess, to the next generation. The result is a change in the genetic composition of the population. This is **evolution**, the basis of which is *a change in allele frequency in the population.*

The classic example of directional selection is provided by the changes in the proportions of light and dark forms of the moth *Biston betularia*, as a result of industrial pollution.

Speciation

Speciation is the evolution of new species from ancestral species. A **species** is defined as a group of organisms with similar morphological, physiological, biochemical and behavioural features that can interbreed to produce fertile offspring, and which is reproductively isolated from other species. New species evolve when genetic differences develop that prevent them from freely interbreeding.

Speciation through isolation: allopatric speciation

Allopatric speciation is the evolution of new species as a result of **geographical isolation**. The process takes place as follows:

- The ancestral species expands its range into new locations, although the different populations are capable of interbreeding. There is regular gene flow in the gene pool.
- In one locality some **physical barrier** isolates a population geographically. This barrier could be a river, mountain range or stretch of ocean. *Gene flow with the ancestral population is prevented.*

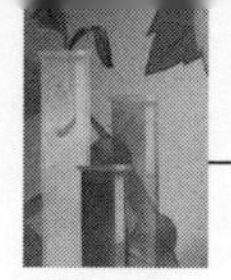

- The isolated population and ancestral population are subjected to *different selection pressures*. For example, different selection pressures may arise because of differences in the food source or in the climate (wetter/drier and warmer/cooler). If the isolated population is small (which often happens on islands), genetic drift (random fluctuations in allele frequencies) and the founder effect (the individuals that became isolated were not genetically representative of the ancestral population) also affect its genetic composition.
- The isolated and ancestral population *diverge genetically* and have, for example, different morphological appearances. If members of the isolated population remain potentially capable of interbreeding with members of the ancestral population then they do not form a different species (though they may be regarded as a subspecies).
- The two populations may *diverge genetically to the extent that interbreeding is prevented*. A new species has evolved. In the speciation of fruit flies (*Drosophila* spp.), genetic differences arose which caused differences in the size of genitalia and mating behaviour.

If the new species and the ancestral species come together again because the barrier is no longer effective, then there will be further interaction. Since they are most likely to have similar niches, there will be interspecific competition with two possible outcomes:

- One species (possibly the new species) will eliminate the other.
- Both species will evolve further and there will be niche divergence (see Figure 35). This is because the most intense competition will occur between the members of both species in the area of niche overlap and so they will be selected against.

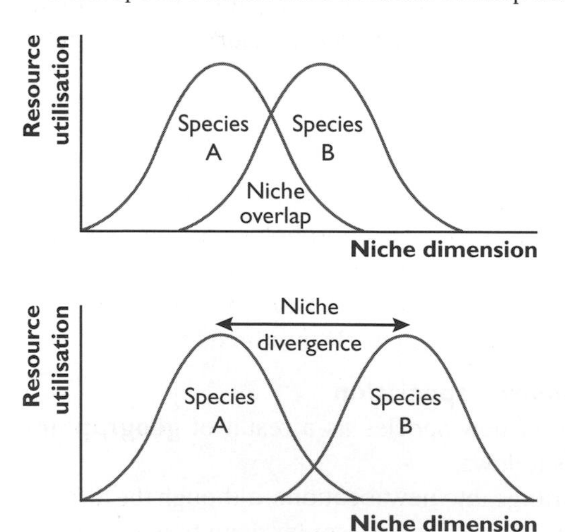

Competition is more intense in the area of niche overlap, resulting in selection against individuals with overlapping characteristics

The niche dimension might represent the 'size of seeds eaten', as in the case of Galapagos finches (where the character in question is beak size) and resource utilisation would be 'amount of seed eaten'

There is further evolutionary change (in the case of Galapagos finches those birds in the range of overlap were selected against) leading to divergence in the niches of the two species.

Figure 35 Niche divergence between two closely related species

The Galapagos finches, observed by Darwin, are a well-known example of speciation and niche divergence. The Galapagos Islands are a group of 19 islands in the Pacific

Ocean colonised by a small group of ancestral finches from South America almost 3 million years ago. On the outer islands, different species evolved to give rise to the 13 species of finch. Two species of finch (*Geospiza fortis* and *G. fuliginosa*) illustrate niche divergence: where they are found separately on different islands their beak size is similar (the size of food eaten is similar) but where they are found together on the same island their beak sizes are quite distinct.

Speciation through hybridisation and polyploidy

Some closely related species may hybridise. However, the offspring are sterile because the chromosomes from two different species cannot pair and so meiosis cannot take place. Therefore, no gametes are produced. In animals, hybrids cannot reproduce at all — mules can only be produced from a cross between a horse and a donkey (not from a cross between mules). In plants, hybrids may reproduce asexually (vegetative propagation) even if the populations formed are slow growing and generally localised. Further, and importantly, sterility in plants may be overcome if interspecific hybrids double their chromosome number to become polyploids, in which case their chromosomes can pair again during meiosis.

Hybridisation, followed by polyploidy, has resulted in the formation of new species of primula (*Primula kewensis*), durum wheat (*Triticum durum*), bread wheat (*Triticum aestivum*) and hemp (*Galeopsis tetrahit*). A new species of cord grass (*Spartina anglica*) has evolved from the hybridisation of a British species (*S. maritima*) and an introduced American species (*S. alterniflora*) and the subsequent polyploidy of the sterile hybrid (*S.* × *townsendii*) (see Figure 36).

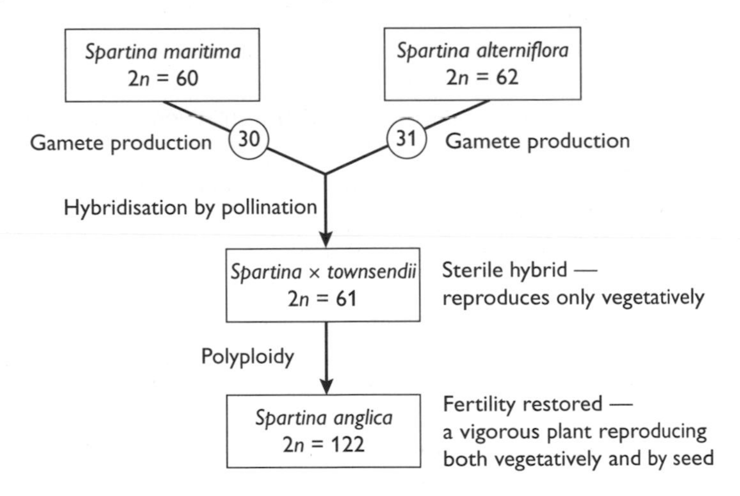

Figure 36 Speciation in* Spartina *by hybridisation and polyploidy

Approximately half of all plant species are polyploids. Polyploids tend to be more vigorous (more genetic information) and bigger (larger cells, since the nuclei contain more chromosomal material). Those with an uneven number of chromosome sets — for example triploids — are sterile. Bananas are triploids and, being sterile, are seedless.

Kingdom Plantae

All plants are multicellular, composed of eukaryotic cells connected by plasmodesmata through cellulose walls. They are photosynthetic and many cells contain chloroplasts. They also have a life cycle exhibiting an alternation of generations, with a *sexually reproducing, haploid* **gametophyte** generation (in humans and most animals only gametes are haploid), and an *asexually reproducing, diploid* **sporophyte** generation. A generalised plant life cycle is shown in Figure 37.

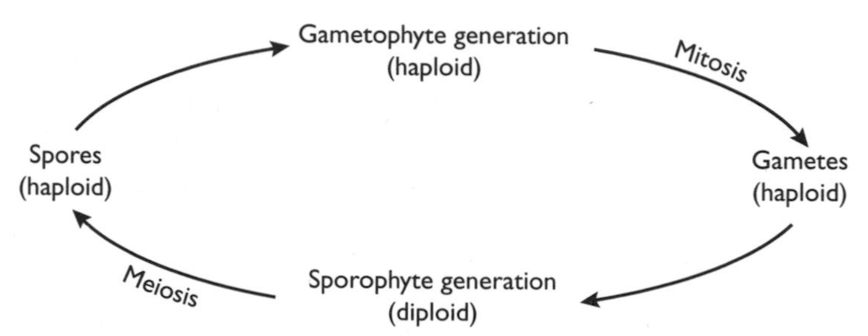

Figure 37 A generalised plant life cycle

A glossary of terms used in describing plant life cycles is shown in Table 13.

Table 13 A glossary of terms used in describing plant life cycles

Term	Definition
Haplophase	The phase in the reproductive cycle after meiosis and before fertilisation in which cells and organisms are haploid
Diplophase	The phase in the reproductive cycle after fertilisation and before meiosis in which cells and organisms are diploid
Gametophyte	The developed haplophase plant is called the gametophyte because it produces gametes (by mitosis)
Gametangium (pl. gametangia)	The organ in which gametes are formed
Antheridium (pl. antheridia)	The gametangium, which produces the male gametes in lower plants (mosses and ferns)
Archegonium (pl. archegonia)	The organ that produces usually one egg in lower plants
Sperm (antherozoids in lower plants)	The male gametes, which are flagellated in lower plants; in flowering plants they are carried by the pollen tubes
Ovum (pl. ova)	The female gamete, egg, to which the male gametes move (or are moved)
Sporophyte	The developed diplophase plant is called the sporophyte because it produces spores
Sporangium	A walled structure in which spores are formed by meiosis
Spores	Haploid cells that germinate to produce the gametophyte

The various divisions that make up the kingdom Plantae exhibit differences in their structure and their life cycles.

Division Bryophyta

Bryophytes include **mosses**.

Structure of a moss

The dominant phase in the life cycle is the **gametophyte** (haploid plant). This is the plant that is recognisable as a moss plant. The moss lacks true roots, stems and leaves, and does not possess vascular tissue.

The gametophyte grows from a spore. It first forms a filamentous green mass called a **protonema**, which matures into the 'leafy' stage. The mature gametophyte has the following features:

- It has leaf-like structures, often just one cell thick, to increase the surface area for photosynthesis.
- It has a stem-like structure that supports the leaf-like structures.
- The lack of vascular tissue and reliance on turgor for support means that mosses are relatively small.
- **Rhizoids**, a group of fine multicellular filaments, anchor the moss to the ground.
- Rhizoids are also used to obtain water and mineral ions, though absorption can occur over the entire gametophyte surface. However, since the rhizoids do not deeply penetrate the soil, mosses are limited to areas where water and ions occur close to the surface.
- The gametophyte lacks a cuticle. For this reason, mosses require a moist environment, though some can survive long periods of dormancy in a dry state.
- At a certain time of the year, **gametangia** develop: **antheridia** for the production of sperm (also called antherozoids) and **archegonia** for the production of eggs (ova).

The **sporophyte** develops from a zygote (fertilised egg) and consists of a seta (stalk) and a capsule, the sporangium, which contains spores. It is totally dependent on the gametophyte for its nutrition, i.e. there is no free-living sporophyte. The sporophyte possesses a cuticle and stomata to control water loss.

Life cycle of a moss

The principle processes in a life cycle are **fertilisation** (also called **syngamy**) and **meiosis**.

The production of gametes and fertilisation

- Archegonia produce eggs, with one egg retained at the base of each archegonium.
- Archegonia and antheridia may develop on different plants, so ensuring cross fertilisation.
- Both eggs and sperm are produced by mitosis and so any produced in the same plant are genetically identical.
- Sperm, produced in antheridia, are flagellated and are released to the exterior.

- Fertilisation is often dependent on water droplets (rain) to splash the sperm from the antheridial 'cups' into the archegonial 'cups'.
- The flagellated sperm swim in a layer of water down the neck (called the venter) of the archegonium towards the egg at its base.
- Fusion of a sperm and an egg produces a diploid zygote.

Meiosis and the production of spores

- The zygote, at the base of the archegonium, develops into a seta topped by a capsule containing a sporangium.
- In the sporangium haploid spores are produced by meiosis and, therefore, are genetically variable.
- When the spores are ripe the capsule splits, allowing the spores to be shaken out in the wind.

The life cycle of a moss is summarised in Figure 38.

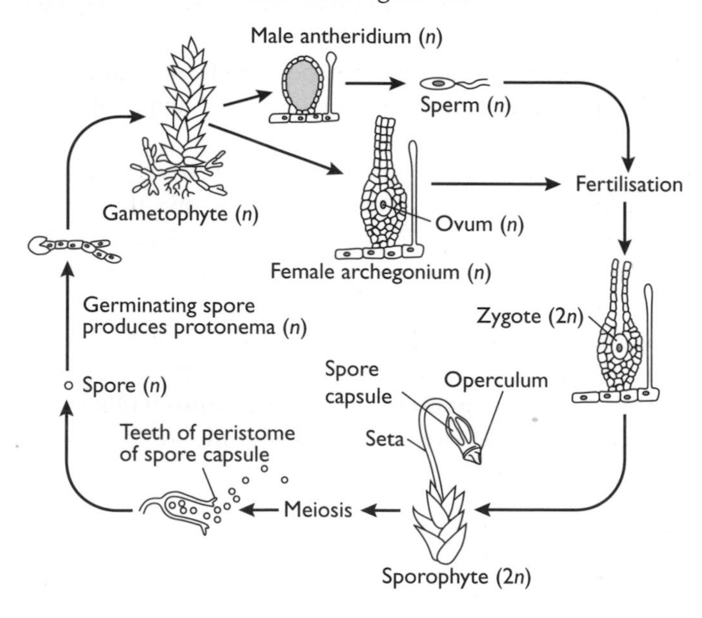

Figure 38 The life cycle of a moss

Division Tracheophyta: subdivision Pteridophyta

The tracheophytes are vascular plants. The division includes the ferns (pteridophytes) and flowering plants (spermatophytes).

Structure of a fern

The dominant phase in the life cycle of a fern is the **sporophyte** (diploid plant). This is the plant that is recognisable as a fern.

The sporophyte develops from a zygote. It matures into a plant that has true roots, a stem (an underground stem called a rhizome) and leaves (which are divided into leaflets, called pinnae). Ferns possess vascular tissue and can be relatively large. The leaves are covered by a cuticle and possess stomata. On the underside of the leaves clusters of sporangia develop. Each cluster of sporangia is called a sorus (pl. sori) and is usually protected by an inverted umbrella-shaped structure called an indusium.

The **gametophyte** develops from a spore. It is a separate plant called a **prothallus**. This flattened, heart-shaped structure bears the gametangia (archegonia and antheridia) on its underside. The prothallus is anchored in the soil by rhizoids though absorption of water and ions takes place over the entire surface.

Life cycle of a fern

Meiosis and the production of spores

- Spores are produced in the sporangia by meiosis.
- The spores are released when the annulus layer (along the edge of the sporangium) dries out and the sporangia break open. They are then dispersed by the wind.
- The spore-producing capacity of a fern sporophyte is truly enormous: each plant has many leaves, each leaf has many leaflets, each leaflet has many sori, each sorus has many sporangia and each sporangium produces many spores.

The production of gametes and fertilisation

- Archegonia and antheridia produce eggs and sperm respectively, by mitosis.
- The sperm are released and swim in a film of water towards and into the archegonia where fertilisation of eggs takes place.

The fern life cycle is summarised in Figure 39.

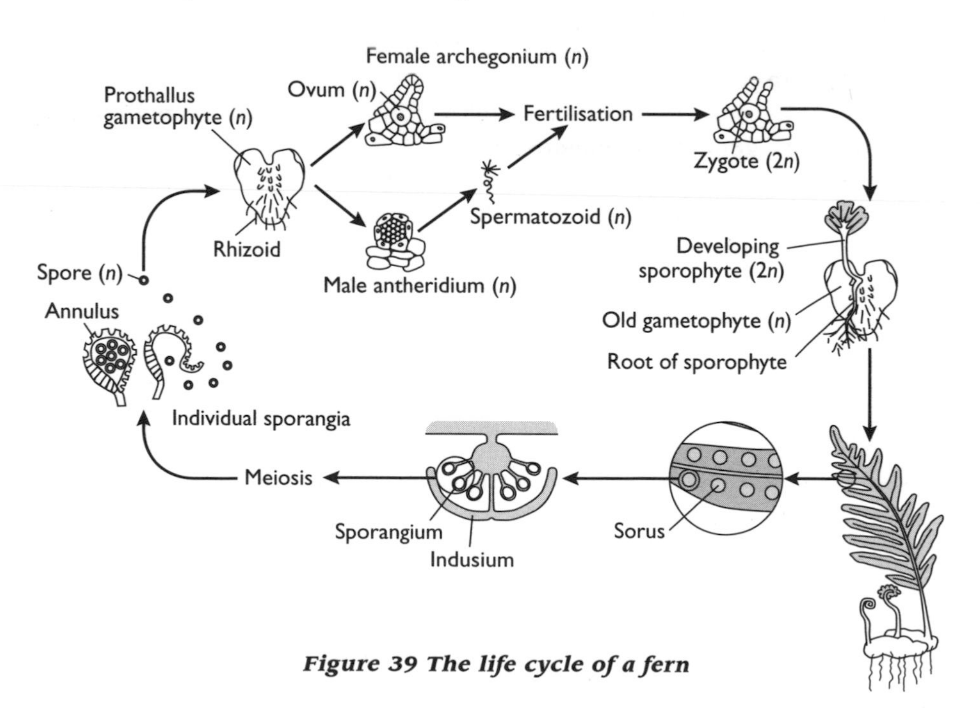

Figure 39 The life cycle of a fern

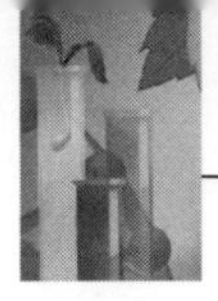

Division Tracheophyta: subdivision Spermatophyta

Spermatophytes are the flowering plants, which have seeds enclosed in an ovary.

Structure of a flowering plant

The sporophyte (diplophase generation) is dominant and the gametophyte (haplophase generation) is restricted to a few mitotic divisions. Flowering plants possess roots (with root hairs for absorption), stems and leaves, and have well developed vascular tissue. The plant produces flowers which contain the sporangia. There are two types of sporangia and so two types of spore. Therefore, a flowering plant is described as **heterosporous**. The **ovary** (in the **carpel**) contains the **ovule** — the **megasporangium**. This produces a **megaspore** (**embryo sac**) in which the female gametophyte produces an egg. The **pollen sacs** (in the **anther**) represent the **microsporangia**. They produce **microspores** (**pollen grains**) in which male gametophytes and male gametes may develop.

Life cycle of a flowering plant

Meiosis and the production of spores

- A diploid cell in the ovule (megasporangium) divides by meiosis to produce four haploid cells, only one of which persists to become the embryo sac (megaspore).
- Diploid cells in the pollen sacs (microsporangia) divide by meiosis to produce haploid cells that become the pollen grains (microspores).

The production of gametes and fertilisation

- The embryo sac undergoes three successive mitotic divisions to produce eight cells: three antipodal cells towards the top, **two polar nuclei** centrally, and **an egg** and two synergids at the base.
- These mitotic divisions and the mature embryo sac formed represent the **female gametophyte** generation.
- The pollen grain nucleus undertakes a mitotic division to produce a **pollen tube nucleus** and a **generative nucleus**.
- If the pollen grain arrives on a receptive stigma it will germinate and, as the pollen tube grows towards the ovule, the generative nucleus divides mitotically to produce **two male gametes**.
- These germinating pollen grains represent the **male gametophyte** generation.
- When the pollen enters the embryo sac the male gametes are released near the egg and the polar nuclei.
- One male nucleus fuses with the egg to produce a diploid **zygote** that develops to form an embryo plant in the seed. This is the true fertilisation.
- The other male gamete fuses with the two polar nuclei to produce a **triploid endosperm** which may develop as a food storage area in the seed. There is said to be a **double fertilisation**.

The flowering plant life cycle is summarised in Figure 40.

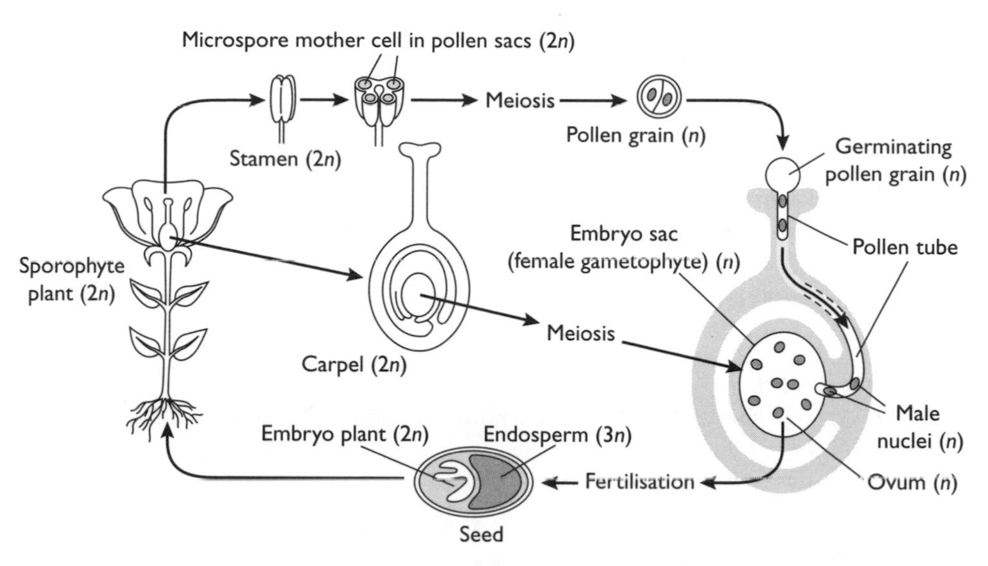

Figure 40 The life cycle of a flowering plant

Comparison of the divisions in the kingdom Plantae

Plants exhibit a life cycle with an alternation of generations in which:

- meiosis produces spores
- mitosis produces the gametes

A comparison of the plant divisions also reveals evolutionary trends in the kingdom Plantae.

There is progression towards greater adaptation for life on land:

- In mosses, the male gametes are released close enough to the eggs to be able to swim towards them in a film of water. The dominant gametophyte relies on turgor for support and lacks a cuticle and stomata.
- Ferns rely on a film of water for the fertilisation process. The dominant sporophyte possesses vascular tissue, true roots to aid water uptake, is covered by a cuticle and has an epidermal layer containing stomata.
- Flowering plants possess vascular tissue, true roots, a cuticle and stomata. They do not rely on water for the fertilisation process since both eggs and male gametes are enclosed (and pollen grains are highly resistant to desiccation).

There is also progression towards dominance of the sporophyte generation:

- in mosses the gametophyte generation is dominant
- in ferns the sporophyte generation is dominant
- in flowering plants the sporophyte generation is completely dominant with the gametophyte apparently degenerate

The sporophyte represents the diplophase and there are advantages in being diploid: two alleles of each gene exist to provide greater vigour, potentially harmful recessive alleles are 'hidden' and there is greater genetic variability.

As a result, flowering plants are the dominant plants on Earth today.

Tip Sometimes examiners might ask you to 'compare and contrast' the life cycles of two plant types. 'Compare and contrast' emphasises the need for both similarities and differences — students often ignore the similarities. Make sure you pay attention to the command term used in the question.

Practical work
Study appropriate living and preserved specimens, prepared slides and photographs.

Kingdom Animalia

All animals are multicellular, composed of eukaryotic cells lacking cell walls. They are diploid: the gamete is the only haploid cell. They are heterotrophs, feeding on other organisms (or their remains) and digesting them in an internal cavity (**enteron**). Most animals move from place to place in search of food. All animals develop from an embryonic stage, a hollow ball of cells called a **blastula**, which forms different germ layers, i.e. layers that give rise to all the animal's tissues and organs.

The various phyla that make up the kingdom Animalia exhibit differences in feeding and in the development of germ layers:

- They may possess a digestive cavity, in which case digestion takes place both in the cavity (extracellular) and in the cells lining the cavity (intracellular), or a through gut where digestion is completed extracellularly.
- They may have two germ layers (**diploblastic**) or three germ layers (**triploblastic**). If triploblastic, a body cavity (**coelom**) may develop.

Phylum Cnidaria

The cnidarians include sea anemones, jellyfish, corals and the freshwater *Hydra*.

Structure of *Hydra*

The body of *Hydra* (see Figure 41) has the following features:

- It is diploblastic, i.e. the body wall is built of two layers of cells. The outer **ectoderm** layer is separated from an inner **endoderm** by jelly-like **mesoglea**.
- The digestive cavity, called the **enteron**, is sac-like.
- The enteron has only one entrance, which, though called the mouth, also serves as the exit for undigested remains.

- The body is **radially symmetrical**, so it can respond to its environment through 360° — for example, *Hydra* can catch prey in any direction.
- The body possesses **tentacles** with stinging cells, called **cnidocysts**, in the ectoderm.
- The body is supported by the aqueous medium and a hydrostatic skeleton formed by the fluid-filled enteron.
- Although often sedentary, *Hydra* can undertake cartwheel movements.

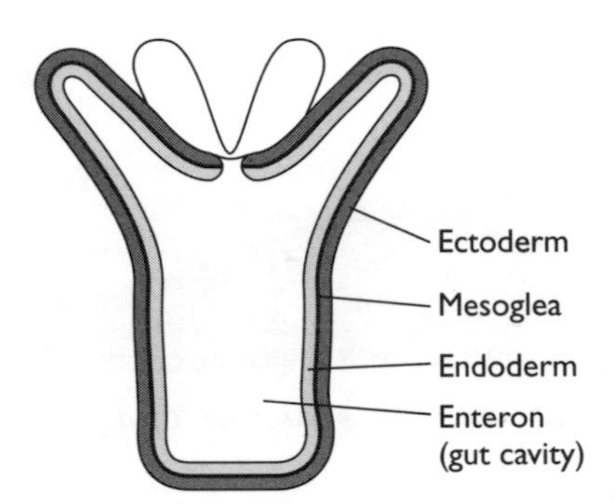

Figure 41 The body structure of Hydra

Feeding in *Hydra*

The process of feeding involves the following:

- *Hydra* predates live prey animals, for example the water flea *Daphnia*.
- When a prey animal touches a cnidocyst, its poison tube with barbs shoots out, impaling the prey.
- The paralysed animal is conveyed to the mouth and into the enteron by the tentacles.
- Endodermal cells secrete enzymes onto the food so that digestion is *initiated* **extracellularly**.
- Partly digested fragments of food are ingested into the endodermal cells by endocytosis. Digestion is *completed* **intracellularly**, by the enzymes in lysosomes.
- Undigested remains of the food are egested from the enteron through the mouth.

Phylum Platyhelminthes

The platyhelminths are commonly called flatworms and include the planarians.

Structure of a planarian

The body of a planarian (see Figure 42) has the following features:

- It is triploblastic, i.e. there are three body layers. The outer ectoderm is separated from an inner endoderm by a cellular layer called the **mesoderm**.
- It does not have a coelom, i.e. planarians are **acoelomate**.
- It is dorso-ventrally flattened and, as a result, has a large surface area-to-volume ratio.

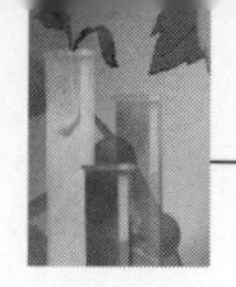

- The gut cavity has only a single opening.
- The gut is branched, permeating all parts of the body.
- As a result of the branched gut and overall flattened shape, no cell in the body is very far from either the gut or the permeable ectoderm. Therefore, flatworms lack a circulatory system.
- The body is **bilaterally symmetrical**.
- The body is supported by the aqueous medium, though much of the mesoderm has a supportive role.
- Flatworms exhibit a gliding movement brought about by the action of cilia on a layer of mucus (produced by slime cells).

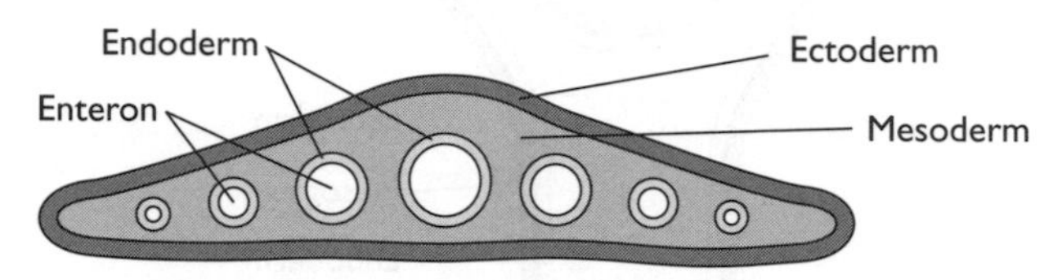

Figure 42 The body structure of a planarian

Feeding in a planarian

The process of feeding involves the following:

- Most planarians are detritivores but there are a few active carnivores.
- Once in contact with food, they evert a pharynx (the muscular section of the gut connecting to the mouth) by which food is taken into the **highly branched gut**.
- Enzymes are secreted onto the food so that digestion is initially extracellular.
- Partly digested fragments of food are ingested into the endodermal cells by endocytosis. Digestion is completed intracellularly by the enzymes in lysosomes.
- The highly divided gut means that transport distances for digested food are reduced.
- Undigested remains of the food are egested from the gut through the mouth.

Phylum Annelida

The annelids are segmented worms and include the earthworm.

Structure of an earthworm

The body of an earthworm (see Figure 43) has the following features:

- It is triploblastic. The outer ectoderm is separated from the inner endoderm by the mesoderm.
- Earthworms are **coelomate**. The body cavity (coelom) is formed in the mesoderm.
- The coelom is large. It is fluid filled, serving as a hydrostatic skeleton and it provides room for the development of a range of organs.
- There is a **through** (one-way) **gut**, with both a mouth and an **anus**.
- The body is bilaterally symmetrical.
- It is **metamerically segmented** (i.e. the body is divided up into many repeating segments, each containing identical copies of nerve ganglia, excretory structures and muscles).

- It is supported by a hydrostatic skeleton provided by segmented, fluid-filled coeloms (body cavities).
- Locomotion relies on the action of muscles acting on the hydrostatic skeleton, aided by external bristles of chitin, called **chaetae**.
- For ease of movement, earthworms are long and thin.

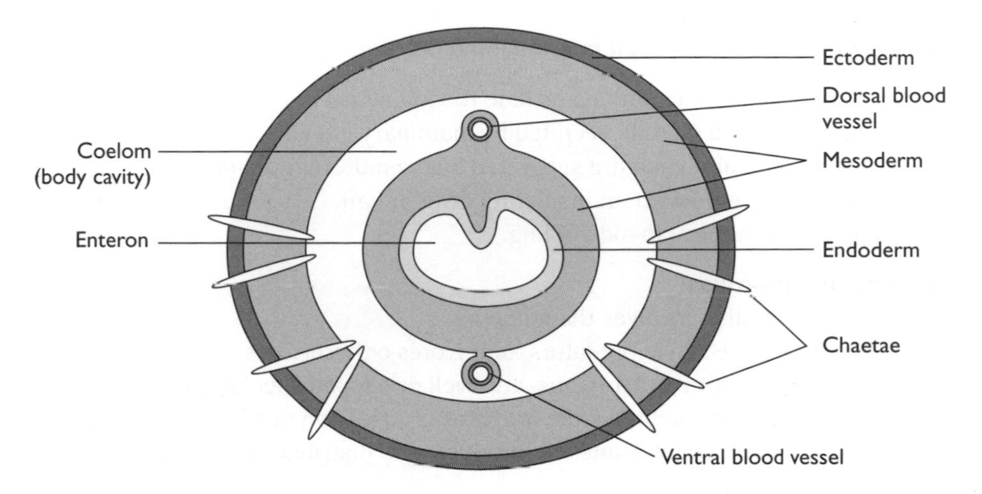

Figure 43 The body structure of an earthworm

Feeding in an earthworm

The process of feeding involves the following:

- Earthworms are detritivores.
- Food is ingested via the mouth.
- The **one-way gut** allows regional specialisation. Food is drawn into a muscular pharynx, passed along an oesophagus to be stored in the crop, slowly released into the muscular gizzard where it is crushed, and then digested and absorbed in the intestine.
- A one-way gut also allows food to be ingested continually while previously ingested food is digested.
- Digestion is completed in the gut, i.e. it is *entirely* **extracellular**.
- The products of digestion are absorbed into, and distributed by, a well developed blood system.
- Undigested remains of the food are egested from the anus and deposited on the surface of the ground as worm casts.

Phylum Chordata

The chordates include the vertebrates — fish, amphibians, reptiles, birds and mammals.

Structure of a mammal

The body of a mammal has the following features:

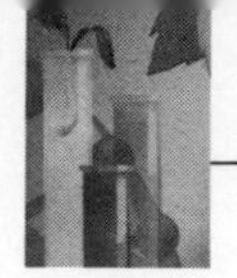

- It is triploblastic, coelomate, bilaterally symmetrical, metamerically segmented, and possesses a one-way gut.
- The body cavity (coelom) contains well differentiated digestive, reproductive, circulatory and excretory systems.
- As in all vertebrates, there is a vertebral column.
- There is a post-anal tail.
- The skeleton consists of an internal jointed system of calcified bones.

The features specific to mammals include the following:

- Young are suckled on milk secreted by mammary glands.
- The thorax and abdomen are separated by the muscular diaphragm.
- The body is covered with an insulating layer of hair.
- Sweat glands in the skin aid cooling.

Feeding in a mammal

The process of feeding involves the following:

- Mammals may be active predators, omnivores or herbivores.
- The gut has a mouth and an anus, and well developed specialised regions.
- This specialisation facilitates the progressive digestion of large food molecules.
- Digestion is extracellular and the products are distributed by a well developed circulatory system.

Evolutionary trends in the kingdom Animalia

In the animal kingdom there are the following evolutionary trends:

- from radial symmetry, which suits sessile organisms, to bilateral symmetry, which is a consequence of directional movement (also producing an anterior and posterior end, and a ventral and dorsal surface)
- from a diploblastic to a triploblastic condition, with the mesoderm forming a variety of new tissues
- from a sac-like gut cavity (enteron) to a one-way through gut, providing the conditions for more efficient feeding and digestion
- towards the coelomate condition, providing a body cavity in which organs can grow while fluid cushions them from damage
- towards the development of an internal blood transport system, necessitated by the appearance of the coelom (outer tissues separated from inner tissues), an increase in size and the possession of an impermeable outer covering, which is better adapted to a terrestrial existence

Practical work

Study appropriate living and preserved specimens, prepared slides and photographs.

Questions & Answers

This section consists of two exemplar papers constructed in the same way as your A2 Unit 2 examination paper. There are questions that assess knowledge and understanding, some that require you to apply your understanding to novel situations and a few that assess your knowledge of practical techniques. There is a variety of question styles. Each paper has a total of 90 marks and you have 2 hours to attempt all the questions.

Following each question, there are answers provided by two students — Candidate A and Candidate B. These are real responses. Candidate A has made mistakes that are often encountered by examiners and the overall performance might be expected to achieve a grade C or D. Candidate B has made fewer mistakes. The overall performance is good and is of grade A or B standard.

Examiner's comments

These are preceded by the icon e. They provide the correct answers and indicate where difficulties for the candidate occurred. Difficulties may include lack of detail, lack of clarity, misconceptions, irrelevance, poor reading of questions and mistaken meanings of examination terms. The comments suggest areas for improvement.

Using this section

You could simply read this section but it is always better to be *active* in developing your examination technique. One way to achieve this would be to:

- try all the questions in Exemplar Paper 1 before looking at candidates' responses or the examiner's comments, allowing yourself 2 hours — remember to follow the suggestions in the introduction
- check your answers against the candidates' responses and the examiner's comments
- use the answers provided in the examiner's comments to mark your paper
- use the candidates' responses and the examiner's comments to check where your own performance might be improved

You should then repeat this for Exemplar Paper 2.

Section A

Question 1

The table below shows the number of chromosomes in the nuclei of the cells of some different animals.

Organism	Number of chromosomes
Cow	60
Cucumber	14
Garlic	32
Maidenhair fern gametophyte	30
Bramley apple	51

(a) Determine the number of chromosomes that there would be in
- **an ovum from a cow**
- **a pollen tube nucleus from a cucumber**
- **an endosperm cell from garlic**
- **a spore of the maidenhair fern** (4 marks)

(b) The Bramley apple is a polyploid. Using the information in the table, identify the evidence for this and explain why the Bramley apple is sterile. (2 marks)

Total: 6 marks

Candidates' answers to Question 1

Candidate A

(a) 30 ✓; 14 ✗; 32 ✗; 30 ✓

Candidate A has two correct answers but has failed to recognise that a pollen tube nucleus is haploid and that an endosperm nucleus is triploid. The candidate scores 2 marks.

(b) A polyploid has more than two sets of chromosomes ✓.

This is sufficient for 1 mark, though the candidate might have deduced that the Bramley apple appears to have three sets of 17 chromosomes.

Candidate B

(a) 30 ✓; 7 ✓; 48 ✓; 15 ✗

(b) The Bramley apple has more than two sets of chromosomes ✓. The total of 51 equates to three sets of 17 chromosomes ✓.

Candidate B has responded well apart from the answer to the number of chromosomes in the spore of a maidenhair fern — both the gametophyte and the spore (from which it develops) are haploid.

Overall, Candidate A scores 3 marks and Candidate B scores 5.

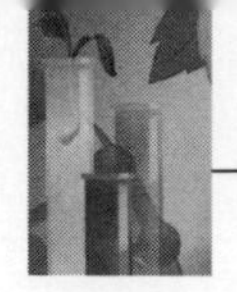

Question 2

The flow diagram below shows four steps (labelled 1 to 4) in glycolysis.

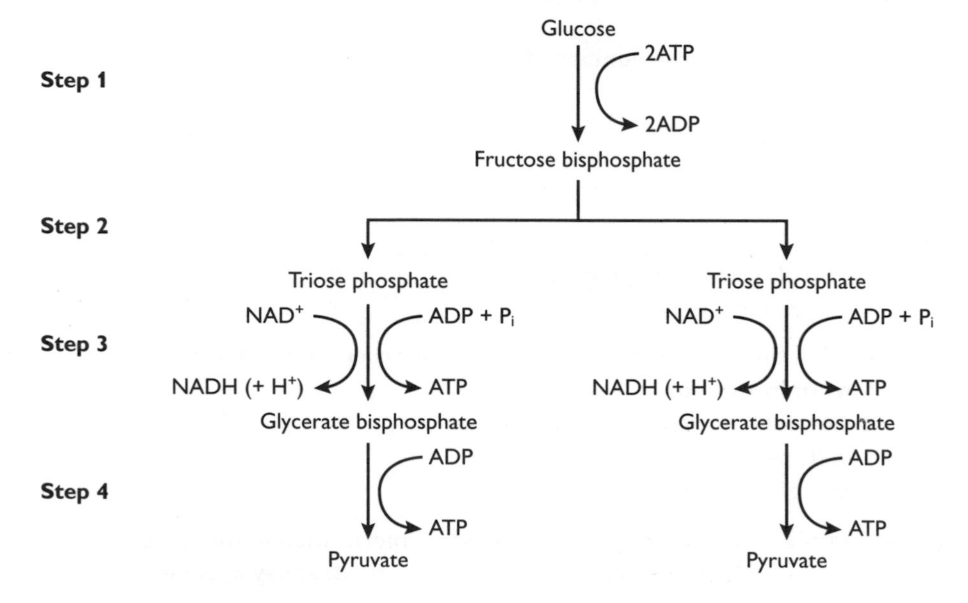

(a) State the net gain of ATP from one molecule of glucose in glycolysis. (1 mark)

(b) Explain why glucose is not normally limiting in a muscle cell. (1 mark)

(c) NAD^+ is required for step 3 and is regenerated by hydrogen being removed from NADH. Explain how hydrogen is removed from NADH when:

(i) oxygen is not limiting (2 marks)

(ii) oxygen is limiting (2 marks)

Total: 6 marks

Candidates' answers to Question 2

Candidate A

(a) 2ATP ✓

This is correct, for 1 mark.

(b) There are large stores of glycogen in the liver ✓.

This is correct, for 1 mark.

(c) (i) Hydrogen is removed in the electron transport chain ✓ where oxygen is the final electron acceptor ✓.

This is sufficient for 2 marks.

(ii) There is no electron acceptor and so NAD^+ is not regenerated and the Krebs cycle stops ✗.

This is irrelevant. The regeneration of NAD^+ occurs via the reduction of pyruvate to lactate.

Candidate B

(a) 2ATP ✓

(b) The liver contains a store of glucose in the form of glycogen ✓.

(c) (i) Hydrogen for NADH passes into the respiratory chain ✓ and oxygen is reduced to water ✓.

(ii) Pyruvate is reduced ✓ to lactate or ethanol ✓ regenerating NAD^+.

The candidate has provided correct answers to each part and scores full marks.

Overall, Candidate A scores 4 marks and Candidate B scores 6.

Question 3

The diagram below represents a transverse section through an earthworm (*Lumbricus*), a member of the phylum Annelida.

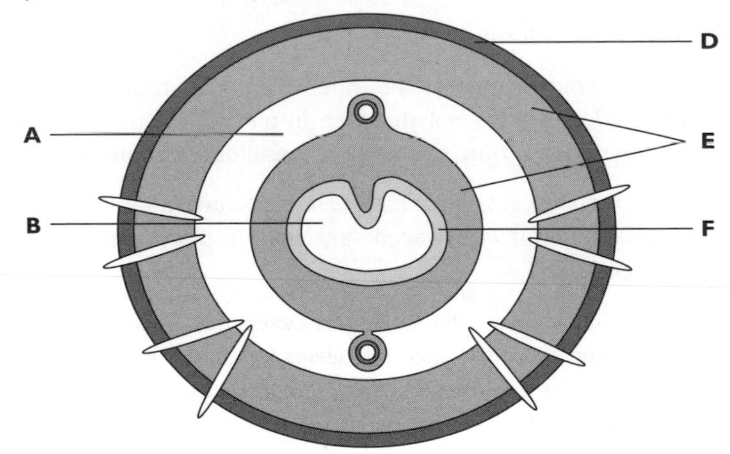

(a) Identify the cavities A and B. (2 marks)

(b) Identify the tissue layers D, E and F. (2 marks)

(c) While annelids possess a blood system, platyhelminths do not. Explain this difference with respect to the body plan of each phylum. (2 marks)

(d) Both annelids and platyhelminths have a digestive system. State two differences between the digestive systems of annelids and platyhelminths. (2 marks)

Total: 8 marks

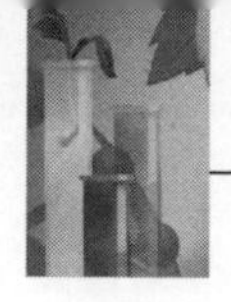

Candidates' answers to Question 3

Candidate A

(a) A is the coelom ✓. B is the gut cavity ✓.

Both are correct, for 2 marks, though the term enteron is more often used for the gut cavity.

(b) D, ectoderm ✓; E, mesoderm ✓; F, endoderm ✓

These three correct answers score both marks.

(c) Platyhelminths are acoelomate and annelids are coelomate ✓.

This observation is correct but does not go far enough. The point is that the coelom separates the inner tissues from the outer tissues (where oxygen is absorbed).

(d) Annelids have a mouth and an anus, platyhelminths only a mouth ✓. In platyhelminths the gut is branched ✓.

These are both correct, for 2 marks. Other correct responses are provided by Candidate B.

Candidate B

(a) A, coelom ✓; B, enteron ✓

(b) D, endoderm ✗; E mesoderm; F, ectoderm ✗

(c) The coelom of annelids separates the inner from outer tissues so a circulatory system is required ✓. This is not the case in platyhelminths since there is no coelom ✓ and the body is thin, so there is a small diffusion distance (✓).

(d) Annelids have a through gut, while platyhelminths have a blind-ending gut ✓. In annelids digestion is entirely extracellular, while in platyhelminths digestion is completed intracellularly ✓.

The candidate has scored fully in all parts except (b) where there has been confusion between ectoderm and endoderm. Although mesoderm is correct, two correct answers are needed here to score 1 mark. It is important to maintain concentration when answering all part-questions, even those that appear straightforward.

Overall, Candidate A scores 7 marks and Candidate B scores 6.

Question 4

The light-dependent stage of photosynthesis takes place on thylakoid membranes in chloroplasts. The diagram below summarises the processes that take place on the thylakoid membrane.

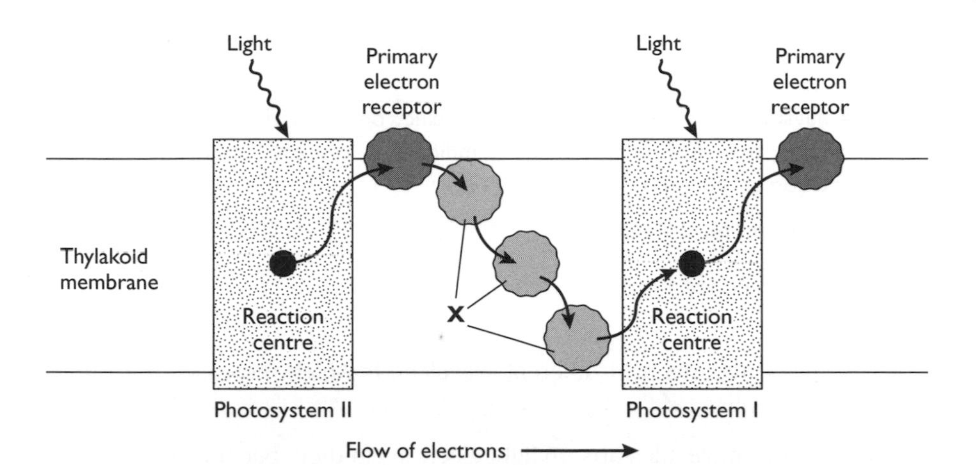

(a) In the photosystems, light is absorbed by different pigments including chlorophyll *a*, chlorophyll *b* and carotenoids. Explain the advantage of the photosystems being composed of different types of pigment. (1 mark)

(b) Explain what happens in the photosystems prior to electrons being emitted from their respective reaction centres. (2 marks)

(c) Each photosystem has the electrons lost due to photoactivation replaced from different sources. State the source of the electrons in each case. (2 marks)

(d) State the name of the molecules labelled X. Explain what happens when electrons pass along these molecules. (2 marks)

(e) Explain what happens when electrons are picked up by the electron acceptor after photosystem I. (1 mark)

Total: 8 marks

Candidates' answers to Question 4

Candidate A

(a) Different pigments absorb at slightly different wavelengths ✓.

Correct, for 1 mark, though Candidate B has a fuller answer.

(b) The energy level of electrons is raised by light absorption ✓.

This is enough for 1 mark but it lacks the detail that the energy is funnelled towards the reaction centre.

(c) PS I — from a photon of light ✗; PS II — from water (OH^-) ✓

The first answer is wrong. The 'lost' electron from PS I is replaced from PS II via the electron carriers. Candidate A scores 1 mark.

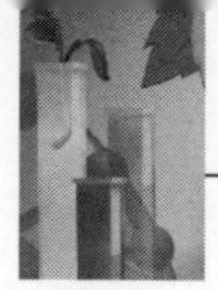

(d) Cytochromes ✓ through which electrons pass at progressively lower energy levels.

The question asks 'what happens when electrons pass along?', to which the answer is that ATP is produced. Candidate A scores 1 mark for naming 'cytochromes' correctly.

(e) Reduces $NADP^+$, forming NADPH ✓

This is correct, for 1 mark.

Candidate B

(a) Each pigment absorbs at a different wavelength which increases the range of wavelengths utilised ✓.

(b) As the pigment molecules absorb light electrons in them become excited ✓. This energy is passed on by resonance to the reaction centre from which a high-energy electron is emitted ✓.

(c) PS I — from PS II ✓; PS II — from hydroxyl (OH^-) ions from the dissociation of water ✓

(d) Cytochromes ✓, each at a progressively lower energy level. As electrons pass along the chain, ATP is synthesised ✓.

(e) Passed on to $NADP^+$, forming NADPH ✓

Candidate B has given correct and full answers throughout, scoring all 8 marks.

Overall, Candidate A scores 5 marks and Candidate B scores 8.

Question 5

The diagram below shows a section of DNA, which includes a desired gene, that has been cut by a genetic engineer from a human chromosome. This gene codes for a medically important hormone.

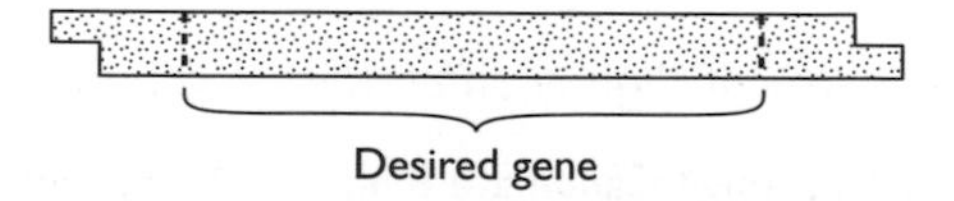

(a) Explain how this section of DNA would have been cut out of the human chromosome. (2 marks)

The diagram below shows a bacterial R-plasmid before, and after, the cut section of DNA was inserted into it. The R-plasmid carries genes conferring resistance to the antibiotics ampicillin and tetracycline.

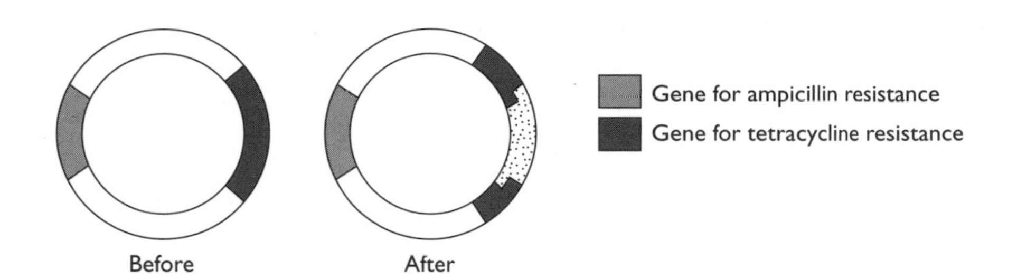

(b) Explain how the section of DNA would have been inserted into the plasmid. (2 marks)

The plasmids are then mixed with bacterial cells (which do not themselves possess plasmids).

(c) Describe how the bacteria may be induced to take up these plasmids. (1 mark)

Only a few plasmids will have incorporated the sections of DNA. Further, only a few of the host bacteria will have taken up a plasmid of any kind. A genetic engineer needs to determine which bacterial cells possess the plasmids with the inserted section of human DNA before cloning them. This is achieved by culturing bacterial colonies on suitable media.

The bacteria are spread onto a nutrient medium. They are then allowed to grow and colonies are then replicated onto plates containing either ampicillin or tetracycline. The bacterial colonies produced on these plates are shown in the diagram below. (The colonies are numbered to allow you to identify them.)

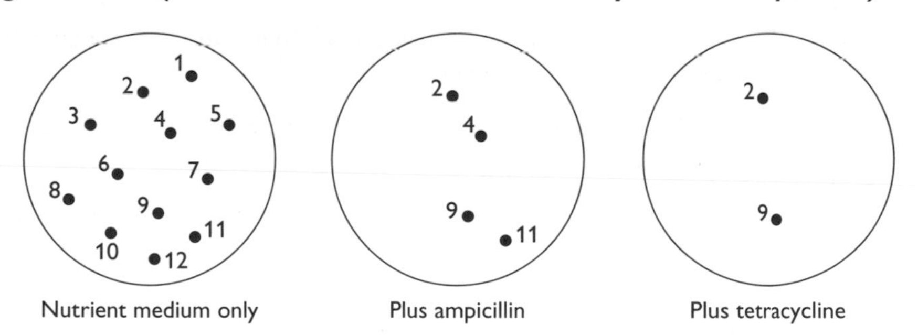

(d) (i) Explain why the agar plate containing nutrient medium only contained the greatest number of bacterial colonies. (1 mark)

(ii) Which colonies contain the bacteria which have taken up the transformed plasmids with the inserted human DNA? Explain your answer. (2 marks)

Total: 8 marks

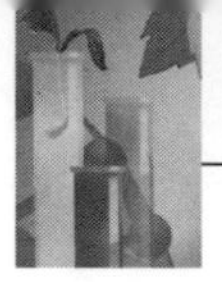

Candidates' answers to Question 5

Candidate A

(a) Using the enzyme restriction endonuclease ✓. It cuts at complementary base sequences ✗.

Candidate A scores 1 mark. Restriction endonuclease cuts at a *specific* base sequence.

(b) The plasmid would have been opened by using a restriction endonuclease ✗ and the section of DNA joined to the open plasmid using the enzyme DNA ligase ✓.

The precise answer is that the *same* restriction endonuclease is used so that the exposed bases (of the sticky ends) are complementary. Candidate A scores 1 mark.

(c) Calcium ions and heat shock ✓

This is correct, for 1 mark.

(d) (i) They don't contain any resistant genes ✗.

This is irrelevant. None of the bacteria will be killed by antibiotic because there is none present in the nutrient medium.

(ii) 4 and 11 ✓. They are resistant to tetracycline but not to ampicillin ✗.

The candidate has identified the colonies correctly, for 1 mark, but the reasoning is confused.

Candidate B

(a) By using a particular restriction endonuclease ✓ that cuts at a specific base sequence ✓ either side of the desired gene (✓).

The bracketed tick indicates that this is a correct point but cannot score because the candidate has already gained the 2 available marks.

(b) By using the same endonuclease enzyme ✓ to cut the plasmid leaving complementary sticky ends ✓ to which the section of DNA is attached using DNA ligase (✓).

(c) Heat shock in the presence of calcium ions ✓.

(d) (i) All the bacteria were capable of growing ✗.

(ii) 4 and 11 ✓. They have taken up the plasmid and so are resistant to ampicillin and the plasmid contains the recombinant DNA as they are not resistant to tetracycline ✓.

The candidate shows good understanding throughout, apart from (d) (i) — all the bacteria were capable of growing because there was no antibiotic in the medium.

Overall, Candidate A scores 4 marks and Candidate B scores 7.

Question 6

An experiment) to investigate the simultaneous inheritance of two characteristics was carried out using *Drosophila melanogaster* (fruit fly).

Female flies with sepia eyes and dumpy wings were crossed with male flies with wild-type (red) eyes and wild-type (long) wings. The F_1 all had red eyes and long wings. These were then interbred to produce an F_2. The results are shown in the table. A chi-squared (χ^2) test was used to determine the probability of the results departing significantly from the expected 9:3:3:1 ratio.

Phenotypes	Observed frequencies	Expected frequencies
Red eye, long wing	73	
Red eye, dumpy wing	19	
Sepia eye, long wing	31	
Sepia eye, dumpy wing	5	

(a) (i) Complete the table to include the expected frequencies. (1 mark)

(ii) State the null hypothesis for the test. (1 mark)

(iii) State the degrees of freedom. (1 mark)

(iv) The χ^2 value was calculated as 4.223. State the probability value for the calculated χ^2 (using the Statistics Sheets). (1 mark)

(v) State your decision about the null hypothesis. (1 mark)

(vi) What can be concluded about the F_1 generation? (1 mark)

(b) State what can be deduced about the inheritance of sepia eye and dumpy wing with respect to the following genetic phenomena:

- **dominance** (1 mark)
- **sex linkage** (1 mark)

(c) State what can be deduced about the genotypes of the original parent *Drosophila*:

- **sepia eyes and dumpy wings female** (1 mark)
- **wild-type (red) eyes and wild-type (long) wings male** (1 marks)

Total: 10 marks

Candidates' answers to Question 6

Candidate A

(a) (i) 72, 24, 24, 8 ✓

(ii) No significant difference in the results ✗

The null hypothesis for this test must specify that there is no significant difference between the observed results and those expected on the basis of theory. The candidate fails to score.

(iii) 1 ✗

The correct answer is the number of classes (4) minus 1, i.e. 3.

(iv) $0.05 > p > 0.01$ ✓

This is correct for one degree of freedom (for which the candidate has already been penalised) and so is given the mark.

(v) The null hypothesis is rejected ✓.

On the basis of the *p*-value given, this is correct and earns the mark.

(vi) They are heterozygous at the genetic loci ✓.

This is correct, for 1 mark.

(b) Both sepia and dumpy are recessive ✓. Neither is sex-linked ✓.

Both answers are correct, for 2 marks.

(c) The parents are heterozygous at one locus but homozygous at the other ✗.

This would not produce dihybrid F_1 flies. Candidate B has the correct answer. Candidate A fails to score.

Candidate B

(a) (i) 72, 24, 24, 8 ✓

(ii) Any difference between the observed results and the expected results is due to chance ✓.

(iii) 3 ✓

(iv) $0.5 > p > 0.1$ ✓

(v) The null hypothesis is accepted ✓.

(ivi) The F1 flies are heterozygous at each of the two loci, e.g. +/se +/dp ✓.

(b) The sepia eye and dumpy wing alleles are both recessive ✓. It is not possible to say if they are sex-linked ✗.

(c) The female parent is se/se dp/dp ✓. The male parent is +/+ +/+ ✓.

The answers are worded well and correct, apart from (b). The alleles cannot be sex-linked. If they had been, the F_1 flies would not all have been wild type and a 9:3:3:1 ratio would not have been produced in the F_2 generation.

Overall, Candidate A scores 6 marks and Candidate B scores 9.

Question 7

Read the passage below and then use the information in the passage, and your own understanding, to answer the questions that follow.

Hereditary haemochromatosis, an inherited disorder of iron metabolism, is one of the most common genetic disorders in Ireland, affecting 1 in every 83 individuals. It is an autosomal recessive disorder. Individuals with haemochromatosis absorb excess iron. The result is iron overload and potential injury to involved organs, including the liver and heart. The symptoms vary among affected individuals, ranging from mild symptoms to life-threatening heart and liver disease.

The *HFE* gene, the gene responsible for hereditary haemochromatosis, was discovered in 1996 when chromosome 6 was analysed as part of the human genome project. This gene encodes for the production of a transmembrane protein, called the HFE protein. The most common mutation of the *HFE* gene is called C282Y. The C282Y mutation replaces the amino acid cysteine with the amino acid tyrosine at position 282 in the HFE protein's amino-acid chain.

Studies of haemochromatosis have progressed through the use of knockout mouse models. Mouse models have been devised to investigate in which organ the gene acts. A mouse model engineered to lack the *HFE* gene in intestinal cells only shows no symptoms of haemochromatosis. However, a mouse model engineered to lack the *HFE* gene in the liver cells only shows all central features of the disease.

It seems that the HFE membrane protein detects the amount of iron in the body. In response to signals from the HFE protein, the liver cells produce the hormone, hepcidin, which is released into the bloodstream and reduces iron uptake in the intestine.

(a) Explain how hereditary haemochromatosis is normally inherited. (1 mark)

(b) Describe what is meant by the 'human genome project'. (1 mark)

(c) With respect to the *HFE* gene, describe what is meant by a 'knockout mouse'. (1 mark)

(d) Explain the C282Y mutation, and the consequence of replacing cysteine with tyrosine in the HFE protein. (2 marks)

(e) Since haemochromatosis involves excessive absorption of iron, it used to be considered a disease of the intestine. However, it is now considered to be a disease of the liver. What is the evidence for this? (2 marks)

(f) Suggest what is happening in a person with haemochromatosis leading to iron overload. (2 marks)

(g) Hereditary haemochromatosis is not usually diagnosed until the age range 40 to 50 years by which time someone with the disease may

already have had children. Suggest what advice a genetic councillor could usefully offer to someone diagnosed with hereditary haemochromatosis. (2 marks)

(h) Other hereditary disorders exist that have serious consequences in early childhood. In such cases, genetic testing for possession of the faulty allele is possible, either of cells from a fetus (pre-natal testing) or of cells from an adult (post-natal testing). While some people may emphasise the advantages of genetic testing, others will argue that such testing is not ethical. Discuss some of the issues involved. (2 marks)

Total: 13 marks

Candidates' answers to Question 7

Candidate A

(a) On the Y chromosome ✗

> There is no evidence that it is confined to males. It is stated in the passage that the gene is located on chromosome 6. Candidate A fails to score.

(b) This was the combined effort of many biologists to sequence all the nucleotides in the human genome ✓.

> This is correct, for 1 mark.

(c) The *HFE* gene has been made inoperative in a mouse for comparison with a mouse with a normal *HFE* gene ✓.

> This is correct, for 1 mark.

(d) The mutation occurred as a result of a mistake in DNA replication ✓.

> This is a correct point, for 1 mark. Other correct points are given in Candidate B's answer.

(e) The disease is linked with the hormone hepcidin, which is produced in the liver ✓.

> This is a relevant point, for 1 mark, but no reference is made to the findings from knockout mouse technology. Candidate B provides a fuller answer.

(f) There is an inability to produce hepcidin ✓.

> This is a relevant suggestion and earns 1 mark. However, the candidate might have noted that this is due to the inability to detect iron levels (since the HFE protein is inoperative).

(g) They should inform their brothers or sisters that they might also have the disease ✓.

> This is a correct point, for 1 mark. An additional piece of advice would be to provide a risk analysis, e.g. that their chance of having the disease would

be 1 in 4, assuming both parents to be heterozygous (the most likely event, especially if they failed to show symptoms of haemochromatosis).

(h) Genetic testing for the disease in a fetus would allow an affected fetus to be aborted ✓.

> This is a relevant point, for 1 mark. There are other issues — for example, some people do not agree with abortion and the knowledge of being a carrier and that future pregnancies might result in abortion is extremely stressful.

Candidate B

(a) Both parents carry the haemochromatosis allele, each passing it on a recessive allele ✓.

(b) The human genome project determined the sequence of all the nucleotides on one set of chromosomes in the human cell ✓.

(c) A knockout mouse (for the *HFE* gene) has the *HFE* gene disabled in order to investigate the gene's role and the consequences for metabolism ✓.

(d) Replacing cysteine with tyrosine alters the tertiary structure of the protein ✓ since cysteine is involved in sulphur bridges ✓. A change in the shape of the protein renders the protein non-functional (✓).

(e) A mouse model with the gene disabled only in intestinal cells shows no symptoms of haemochromatosis ✓. A mouse model with the gene disabled only in the liver cells has the disease ✓.

(f) Too much iron is absorbed in the liver ✗.

(g) Children should be advised that they have the allele but would only be affected if the other parent was a carrier ✓. They may be advised to see a doctor about genetic testing ✓.

(h) Embryos may be produced by IVF and tested so that only healthy embryos are implanted ✓. This is called pre-implantation genetic diagnosis.

> The answers provided are correct, apart from (f) while (h) is incomplete. In (f), the candidate might have noted that 'lacking the HFE protein means that iron levels cannot be detected, so hepcidin is not produced (and so iron uptake is not controlled)'. Other relevant points in (h) have been discussed in Candidate A's answer.

> **Overall, Candidate A scores 7 marks and Candidate B scores 10.**

Question 8

The bracken fern (*Pteridium aquilinum*) grows in dense stands that rarely contain other plants. It is thought that phytotoxins leached by rain from the fern leaves are largely responsible for this suppression.

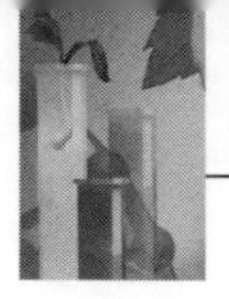

(a) The effects of bracken material on seed germination and seedling growth on three species of coniferous tree, Douglas fir (*Pseudotsuga menziesii*), Scots pine (*Pinus sylvestris*) and Sitka spruce (*Picea sitchensis*), was investigated. In field experiments, water-soluble extracts of bracken leaves were used to irrigate plots of seeds. In studies on seedling growth, other plots had bracken litter incorporated. In both cases, control plots had no bracken material added. The results are shown in the graphs below.

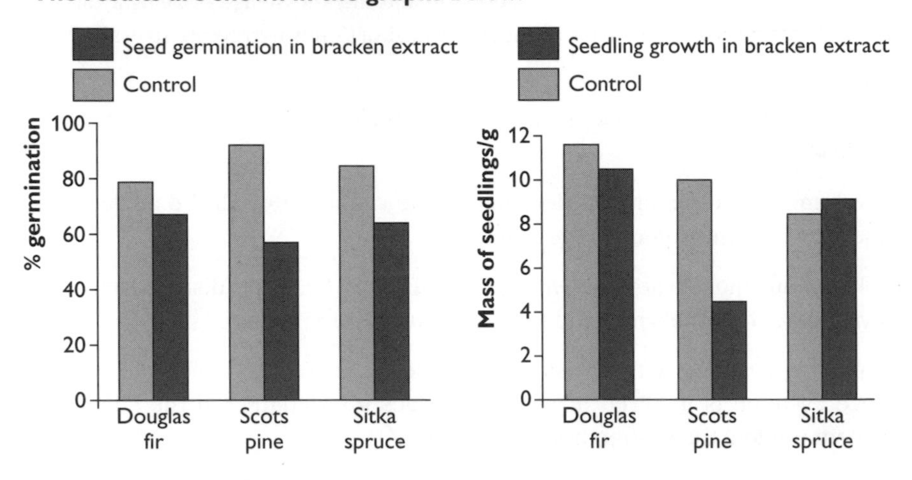

What conclusions can be made about the effect of bracken on the seed germination and seedling growth in the three species of coniferous tree? (4 marks)

(b) Phytotoxins could be contained in any part of the bracken: the leaves, the stem and the roots. Design a laboratory experiment to test which of these parts, if any, exhibited the most inhibitory influence on seed germination. Use lettuce seeds as the test organism. (3 marks)

A number of chemical compounds, including cyanide, have been implicated in the various toxic properties of bracken to animals. Nevertheless, there are insects that eat bracken. These include larvae of sawfly species that feed on bracken leaves.

(c) Wood ants have been shown to impose high rates of predation on larvae venturing onto the leaves. However, when sawfly larvae are attacked they respond by secreting a fluid that repels the ants. Suggest an explanation for the ability of sawfly larvae to repel foraging ants. (2 marks)

Bracken, in common with other ferns, has both a sporophyte and a gametophyte generation.

(d) Give four distinct differences between the fern sporophyte and gametophyte. (4 marks)

Total: 13 marks

Candidates' answers to Question 8

Candidate A

(a) Bracken extracts have a negative affect on seed germination ✓. Seedling growth is negatively affected in Scots pine ✓.

> There are two correct responses, though the candidate does not make any conclusions about the relative effects on the different tree species. Candidate A scores 2 marks.

(b) Remove leaves, stems and roots from a bracken plant and use a mortar and pestle to grind up the parts separately ✓. Add the extract to plots of lettuce seeds. Compare the percentage germination of the seeds treated with extracts from different bracken parts ✓.

> There is no control of variables (e.g. equal mass of plant parts) and no control experiment (lettuce seeds with just water added, i.e. lacking extract). Candidate A scores 2 marks.

(c) The ants are afraid of the fluid secreted by the larvae ✗.

> This answer lacks any detail and the candidate fails to score. Candidate B provides a correct answer.

(d) The sporophyte consists of leaves, stem and roots; the gametophyte consists of a prothallus ✓. The sporophyte has xylem and phloem whereas the gametophyte does not ✓. The sporophyte has a waxy cuticle whereas the gametophyte does not ✓. The sporophyte protects the gametes whereas the gametophyte does not ✗.

> There are three correct points, for 3 marks.

Candidate B

(a) Bracken extracts inhibit seed germination ✓, particularly of Scots pine, ✓ while the effect on Douglas fir is minimal ✓. Seedling growth is reduced in Scots pine ✓ and to a lesser effect in Douglas fir. There is no inhibitory effect on seedling growth in Sitka spruce (✓) — if anything, bracken extract stimulates growth.

> The candidate scores all 4 marks.

(b) Leaves, stem and roots are removed from the fern and equal amounts ✓ are ground separately to make extracts ✓. Lettuce seeds are planted in four different plots. One plot has no bracken solution added as it is used for comparison with the other plots, which are each treated with extract from a bracken part ✓. Percentage germination is compared (✓). Least germination indicates the bracken part that has the most inhibitory effect.

> This full answer scores all 3 marks.

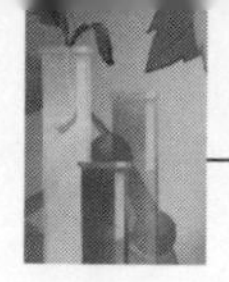

(c) Sawfly larvae must be able to store the bracken toxins ✓ in such a way that the larvae's metabolism is unaffected. The toxin in the fluid released by the larvae when they are attacked is detected by the ants ✓, which are repelled.

These are both good suggestions. Candidate B scores 2 marks.

(d) The sporophyte is the dominant generation ✓. The sporophyte contains sporangia whereas the gametophyte contains gametangia ✓. The sporangia produce spores by mitosis ✗ whereas the gametangia produce gametes by meiosis ✗.

Candidate B scores full marks in parts (a), (b) and (c), In part (d), there are only two correct points, for 2 marks. Other correct points are provided by Candidate A.

Overall, Candidate A scores 7 marks and Candidate B scores 11.

Section A total for Candidate A: 43 marks out of 72

Section A total for Candidate B: 62 marks out of 72

Section B

Quality of written communication is awarded a maximum of 2 marks in this section. (2 marks)

Question 9

Write an essay on how evolutionary change may take place in populations and how this could lead to speciation. (16 marks)

Total: 18 marks

Candidates' answers to Question 9

Candidate A

Natural selection acts on the genetic variation of a population ✓. Genetic variation results from meiosis. This is because of crossing over and the independent assortment of chromosomes. Crossing over happens when chiasmata form during prophase I of meiosis. Independent assortment happens because of the random alignment of homologous chromosomes during metaphase I. Both stabilising selection ✗ and directional selection cause evolutionary change ✓. The type that is selected survives and is favoured ✓. If the environment changes, a type, not previously favoured, may be selected ✓. This type will reproduce and pass on more genes to the next generation ✓ so there will be an increase in allele frequency ✓.

Speciation occurs when populations become geographically separated ✓. Two species are formed when they can no longer interbreed ✓.

Too much of this answer is not relevant, i.e. the detail on genetic variation. There is confusion about the effect of stabilising selection and, in general, there is insufficient detail, particularly with respect to speciation. Nevertheless, there are eight correct appropriate points. However, since so much of the content is irrelevant there is a lack of coherence and only 1 mark is awarded for QWC.

Candidate B

If the members of a population were the same evolution would not be possible. However, several processes occur to ensure that populations are genetically variable ✓. Natural selection acts on this variation ✓. For example, a population of plants may be adapted differently to the water content of the soil. If conditions are in general moist and the environment stays the same, then plants adapted to moist conditions will be favoured. However, if the environment changes and conditions become drier ✓ then those variants adapted to drier conditions — and not previously favoured — would be selected for ✓. This is called directional selection ✓. This means that they will be more likely to survive and reproduce ✓, so the frequency of 'dry' alleles will increase in the population ✓. This change in allele frequency represents evolution ✓. The end result is that the population stays adapted to its environment ✓.

In the theory of allopatric speciation populations become geographically isolated ✓. For example, some plants may, somehow, have been moved to an island. The island may be wetter: that is, the environmental conditions differ ✓. This means that selection pressures in the original and new environments differ ✓. The result is a divergence of the gene pools of the two populations ✓. If this takes place over a long enough time sufficient genetic differences may occur ✓ to make the populations reproductively incompatible ✓. If the two populations are unable to breed they are known as different species ✓ and speciation has occurred.

The candidate receives the maximum 16 marks for the content. This is a well-structured account and ideas are expressed fluently, so 2 marks are awarded for QWC.

Overall, Candidate A scores 9 marks and Candidate B scores 18.

Section B total for Candidate A: 9 marks out of 18

Section B total for Candidate B: 18 marks out of 18

Paper total for Candidate A: 52 marks out of 90

Paper total for Candidate B: 80 marks out of 90

Section A

Question 1

The diagram below illustrates the life cycle of a common moss or bryophyte.

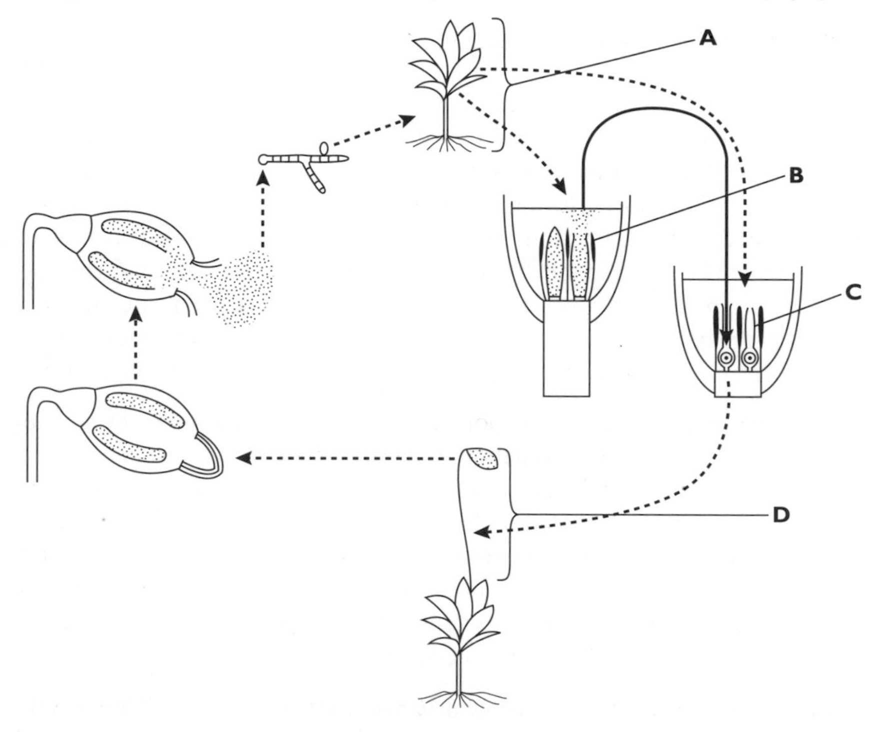

(a) Identify the structures labelled A to D. (4 marks)

(b) Which stage in the moss life cycle is the dominant generation? (1 mark)

(c) During which stage in the life cycle does the process of meiosis take place? (1 mark)

Total: 6 marks

Candidates' answers to Question 1

Candidate A

(a) A, gametophyte ✓; B, anther ✗; C, ovary ✗; D, sporophyte ✓

Anther and ovary are not appropriate terms for the life cycle of a moss. The candidate scores 2 marks.

(b) The gametophyte ✓

This is correct, for 1 mark.

(c) The production of spores in the sporangium ✓

This is correct, for 1 mark.

Candidate B

(a) A, gametophyte ✓; B, antheridium ✓; C, archegonium ✓; D, sporangium ✗

Feature D is the sporophyte. The sporangium is only the top portion of the labelled structure.

(b) The gametophyte generation ✓

(c) In the production of gametes ✗

This is incorrect since mitosis produces the gametes.

Both candidates score 4 marks.

Question 2

The diagram below represents the transfer of electrons from reduced NAD^+ along the respiratory chain. Cytochrome *a* is the final carrier in the chain shown.

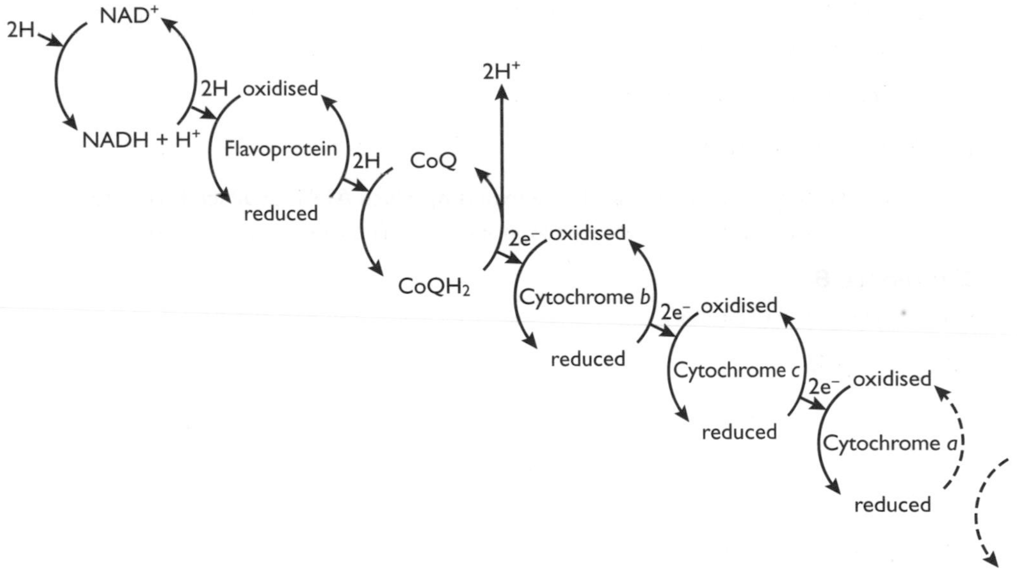

2H
NAD^+
$NADH + H^+$
2H
oxidised
Flavoprotein
reduced
2H
CoQ
$CoQH_2$
$2H^+$
$2e^-$
oxidised
Cytochrome *b*
reduced
$2e^-$
oxidised
Cytochrome *c*
reduced
$2e^-$
oxidised
Cytochrome *a*
reduced

(a) Where, precisely, in a cell are the carriers of the respiratory chain found? (1 mark)

(b) During which series of reactions in the respiration of glucose is most hydrogen made available for the reduction of NAD^+? (1 mark)

(c) In the final step of the respiratory chain, the reduced form of cytochrome *a* is reoxidised. Describe how this occurs. (2 marks)

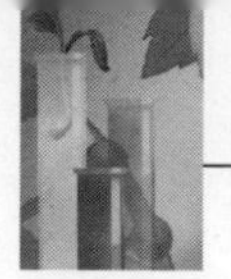

(d) The transfer of electrons along the respiratory chain makes energy available to produce ATP. Describe exactly how this takes place. (2 marks)

(e) What term is used to describe the production of ATP in this process? (1 mark)

Total: 7 marks

Candidates' answers to Question 2

Candidate A

(a) In the mitochondrion ✗

> The precise location (and the question asks for this!) is the cristae of the mitochondrion.

(b) The electron transport chain ✗

> The electron transport chain requires hydrogen. It is the Krebs cycle that makes it available. The candidate fails to score.

(c) Hydrogen is removed ✗ and reduces oxygen ✓.

> It is not hydrogen that is removed but electrons. Candidate A scores 1 mark.

(d) Electrons pass from carriers at a high energy level to carriers at a lower energy level ✓. At particular points in the chain the energy available is enough for an ATP to be produced ✓.

> This is correct, for 2 marks.

(e) Electron transport chain ✗

> The electron transport chain is the means by which ATP is produced. The term for its production is oxidative phosphorylation. The candidate fails to score.

Candidate B

(a) Mitochondrial cristae ✓

(b) Krebs cycle ✓

(c) Electrons from the cytochrome *a* ✓ along with H^+ ions reduce oxygen ✓ forming water.

(d) The electron carriers are at progressively lower energy levels ✓.

(e) Oxidative phosphorylation ✓

> This is a good answer, which almost scores full marks. The response to part (d) is too brief.

> **Overall, Candidate A scores 3 marks and Candidate B scores 6.**

Question 3

The bacterium *Bacillus thuringiensis* possesses a gene (*Bt* gene) that codes for the production of a substance which is toxic to specific insect larvae such as caterpillars. This substance is known as a T toxin. Genetic engineers have succeeded in isolating the *Bt* gene and inserting it into tomato plants. To achieve this, the gene is first inserted into the Ti plasmid from the bacterium *Agrobacterium tumefaciens*. This transformed plasmid can be made to invade tomato cell protoplasts from which the cell wall has been removed enzymatically. Such protoplasts naturally reform their cell walls and can be induced to divide into a callus (mass of undifferentiated tissue) from which new plants may form.

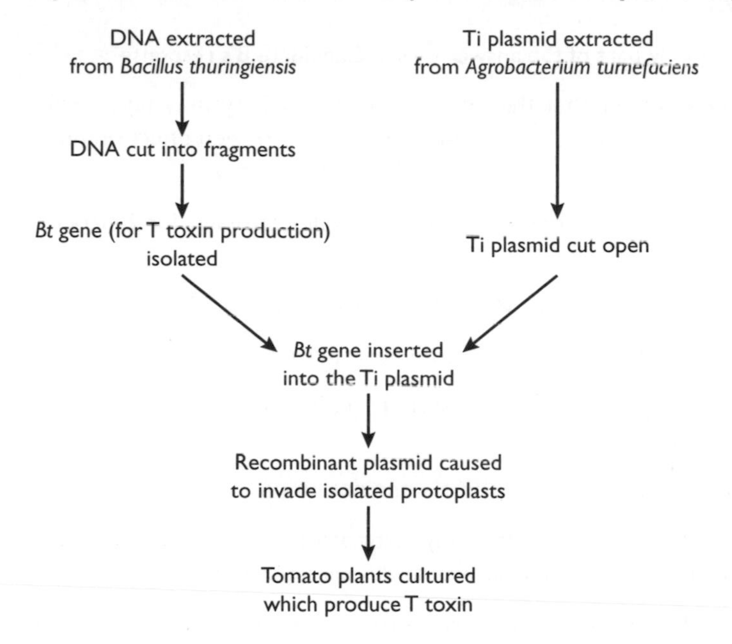

(a) Name the enzymes used in the above procedure to:
- **cut the bacterial DNA into fragments**
- **join the DNA fragment into the opened plasmid**
- **remove cell walls from plant cells** (3 marks)

(b) Identify the vector used in the above genetic engineering procedure. (1 mark)

(c) Explain why it is advantageous to have tomato plants that are genetically modified to produce T toxin. (2 marks)

(d) Suggest two possible problems that may be associated with growing these genetically modified tomato plants. (2 marks)

Total: 8 marks

Candidates' answers to Question 3

Candidate A

(a) Restriction enzyme ✓; DNA ligase ✓; cellulose ✗

Cellulose is the material from which the wall is made — not the enzyme, which is cellulase. Candidate A scores 2 marks.

(b) A plasmid ✗

The vector is not just any plasmid, it is specifically the Ti plasmid. The candidate fails to score.

(c) The plants will contain the toxin, which will kill any caterpillars feeding on them ✓.

This is only part of the answer — see Candidate B's response.

(d) There is a danger that the gene for insect toxicity may be transferred to wild populations ✓. The tomatoes may have an unforeseen effect on human health — for example, they may produce an allergic reaction ✓.

These two suggestions are correct and well-expressed, for 2 marks.

Candidate B

(a) Restriction endonuclease ✓; DNA ligase ✓; cellulase ✓

(b) The Ti plasmid ✓ of *Agrobacterium tumefaciens*

(c) The modified plants will be resistant to insects ✓ and so fewer insecticides need to be used ✓.

(d) There are both moral and ethical considerations.

This answer is not sufficiently explanatory. The candidate fails to score. Candidate A has a good answer to this part-question.

Overall, Candidate A scores 5 marks and Candidate B scores 6.

Question 4

The experiments below refer to changes in the concentration of two intermediates of the light-independent stage of photosynthesis. The plants used were algae and the two compounds measured were ribulose bisphosphate (RuBP) and glycerate phosphate (GP).

(a) In one experiment, the algae were exposed to a period of light followed by a period of darkness. The relative concentrations of ribulose bisphosphate and glycerate phosphate during the experiment are shown in the graph below.

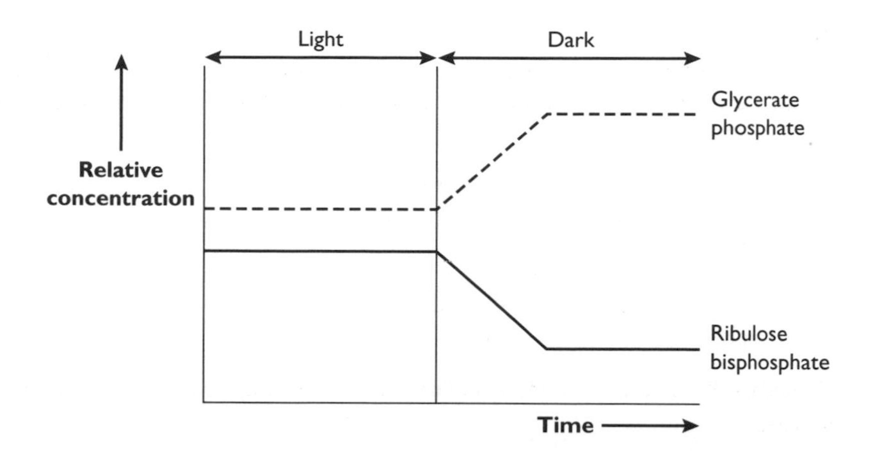

With reference to key reactions in the light-independent stage of photosynthesis, explain the changes in the relative concentrations of the two compounds, glycerate phosphate and ribulose bisphosphate. (4 marks)

(b) In a second experiment, the algae were deprived of carbon dioxide for some time. Following this period of deprivation, carbon dioxide was made available for a further period of time. (Light was available throughout the experiment.)

In this experiment, what would you expect to happen to the relative concentrations of ribulose bisphosphate and glycerate phosphate when carbon dioxide was made available? In each case, explain your answer. (2 marks)

Total: 6 marks

Candidates' answers to Question 4

Candidate A

(a) When light is not available glycerate phosphate cannot be converted to triose phosphate and so its level rises ✓. This is because in the presence of light the NADPH and ATP necessary for the reduction of glycerate phosphate are produced ✓.

The level of ribulose bisphosphate falls in the dark since it cannot be regenerated from triose phosphate ✓ and it is still being converted to glycerate phosphate ✓.

The candidate shows good understanding and scores all 4 marks.

(b) Ribulose bisphosphate would fall ✗; glycerate phosphate would rise ✗.

The candidate has not 'explained' the answer, which is unfortunate since it is a topic that seems to be well known and marks are 'lost' by not reading the question carefully enough. The candidate fails to score.

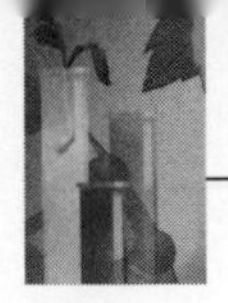

Candidate B

(a) Glycerate phosphate accumulates since light is required to convert it to triose phosphate ✓. The accumulation is also due to it still being produced, from ribulose bisphosphate ✓.

Ribulose bisphosphate concentration drops since it is being converted to glycerate phosphate ✓ and is not being regenerated ✓.

(b) Ribulose bisphosphate decreases as carbon dioxide becomes available and it forms glycerate phosphate ✓. Glycerate phosphate increases since it is being produced in the carboxylation step ✓.

Candidate B has a good understanding of the Calvin cycle and gains the full 6 marks.

Overall, Candidate A scores 4 marks and Candidate B scores 6.

Question 5

(a) The process of polypeptide synthesis involves the transcription of DNA into mRNA and the subsequent translation of mRNA.

(i) State two ways in which mRNA differs in structure from DNA. (2 marks)

(ii) Suggest one reason why mRNA is the template for translation rather than the DNA molecule itself. (1 mark)

(b) The table below shows part of the genetic code. The position of each base in an mRNA codon may be read from the table to denote the amino acid for which it codes. (Abbreviations rather then full names of amino acids are given.)

		Second base					
		U	**C**	**A**	**G**		
First base	U	Phe	Ser	Tyr	Cys	C	**Third base**
		Leu	Ser	–	Trp	G	
	C	Leu	Pro	His	Arg	C	
		Leu	Pro	Gln	Arg	G	
	A	Ileu	Thr	Asn	Ser	C	
		Met	Thr	Lys	Arg	G	
	G	Val	Ala	Asp	Gly	C	
		Val	Ala	Glu	Gly	G	

Using the information in the table, determine:

(i) the amino acid coded for by the codon ACG (1 mark)

(ii) the amino acid transferred by the tRNA with anticodon GUC (1 mark)

(iii) the effect of a mutation changing the codon from UGC to UGG.

Describe what effect this mutation might have on the structure of the protein produced. (2 marks)

(c) The diagram below shows a stage in the synthesis of a polypeptide.

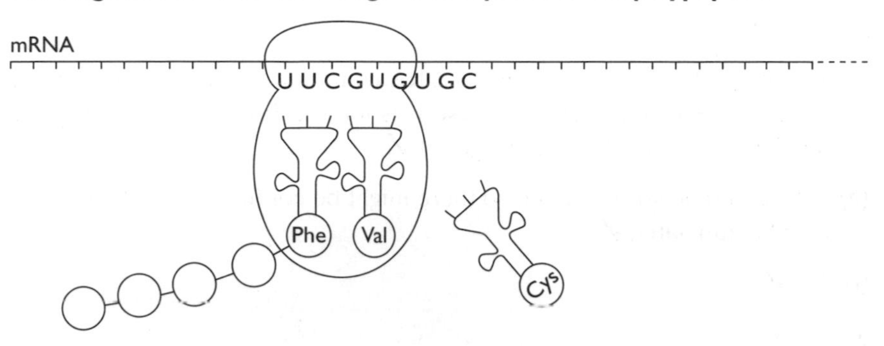

Describe, precisely, what happens at the next stage in the synthesis of the polypeptide. (4 marks)

Total: 11 marks

Candidates' answers to Question 5

Candidate A

(a) (i) Messenger RNA is single stranded, DNA is double stranded ✓. In mRNA thymine is replaced by urine ✗.

Urine is not acceptable for uracil. Candidate A scores 1 mark.

(ii) As the mRNA codons are complementary to the tRNA anticodons ✗.

This is not the reason for DNA not being used for translation. Translation takes place in the cytoplasm, while DNA is confined to the nucleus.

(b) (i) Thr ✓

This is correct, for 1 mark.

(ii) Gln ✓

This is correct, for 1 mark.

(iii) It would code for Trp instead of Cys ✓. The protein would be made up of different amino acids ✗.

The candidate understands the genetic code but has failed to explain the effect of a mutation in altering a protein's tertiary structure and functionality. Candidate A scores 1 mark.

(c) Once the tRNA in the P-site has formed a peptide bond with its amino acid ✗, the tRNA goes off into the cytoplasm to pick up another amino acid ✓. The tRNA in

the A-site moves to the P-site ✓ leaving the A-site empty for the next tRNA with its amino acid to move in ✓.

Peptide bonds form between neighbouring amino acids, not between an amino acid and its tRNA. The candidate gains 3 marks.

Candidate B

(a) (i) Messenger RNA contains ribose, DNA contains deoxyribose ✓. Messenger RNA is smaller than DNA because it is formed from a single gene; DNA is much longer, containing many genes ✓.

(ii) DNA is double stranded and so there might be confusion about which strand is to be translated ✓.

(b) (i) Thr ✓

(ii) Gln ✓

(iii) The amino acid would change from Cys to Val ✗. This would mean the loss of a disulphide bond and a distortion of the tertiary structure ✓.

(c) Once the amino acid at the P-site forms a peptide bond with the amino acid at the A-site (Phe to Val) ✓ its tRNA is released into the cytoplasm to pick up another amino acid ✓. The ribosome moves along one codon length so that the tRNA–amino acid at the A-site is transferred to the P-site ✓ and the A-site becomes available for another tRNA ✓ carrying a specific amino acid (✓) to move in according to the base-pairing relationship between codon and anticodon (✓).

Candidate B has performed very well except in part (b) (ii) where there has been a 'slip' in reading the genetic code table — the first and second bases have been confused.

Overall, Candidate A scores 7 marks and Candidate B scores 10.

Question 6

An orchard contained four varieties of apple tree. To compare fruit production for the four varieties, 30 apples of each variety were chosen randomly and weighed.

From the data, the mean and standard deviation of the mean was calculated for each variety and, for three of the varieties, the 95% confidence limits were determined. The means and 95% confidence limits are shown in the graph below.

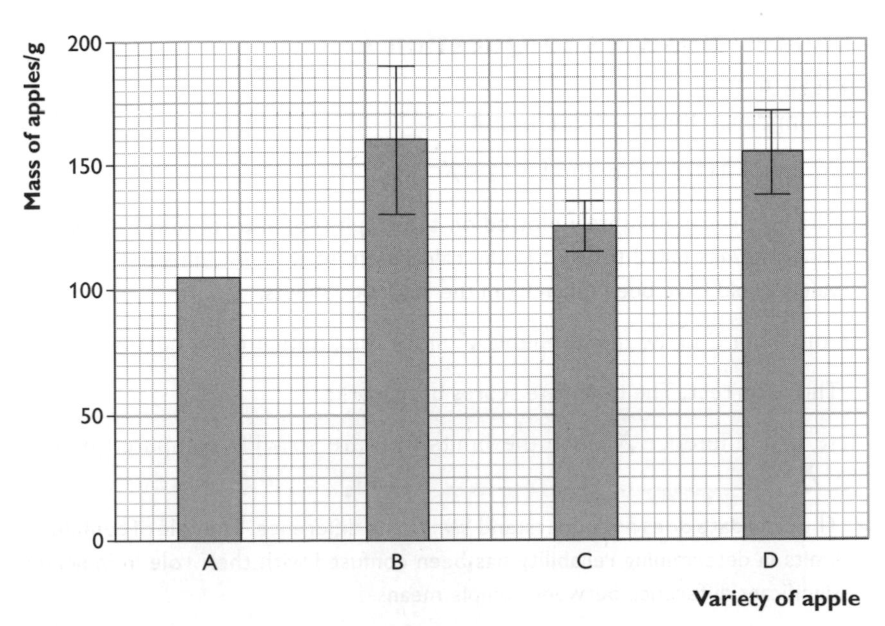

(a) **For the variety A sample (of 30 apples), the mean was 105, while the standard deviation of the mean $\sigma_{\bar{x}}$ was 12.2. (Units are grams.)**

Calculate the 95% confidence limits and plot these on the graph (3 marks)

(b) **Identify which of the means of the four apple varieties is the most reliable. Explain your reasoning.** (2 marks)

(c) **Compare the means and confidence limits for varieties B and C. Do these suggest that there is a significant difference? Explain your answer.** (2 marks)

(d) **A *t*-test was undertaken to compare the mean mass of apples of varieties A and D. The calculated t value was 3.388.**

(i) **State the null hypothesis for the test.** (1 mark)

(ii) **State the degrees of freedom.** (1 mark)

(iii) **State the probability value for the calculated *t* value (using the Statistics Sheets).** (1 mark)

(iv) **State your decision about the null hypothesis.** (1 mark)

(e) **Variety D is sterile and is thought to be a triploid strain of variety A. What other evidence is there that variety D is a triploid strain of variety A?** (1 mark)

Total: 12 marks

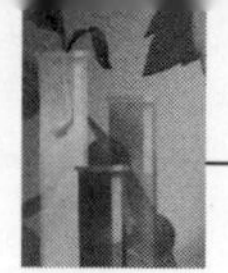

Candidates' answers to Question 6

Candidate A

(a) 95% confidence limits = 105 ± 12.2 ✗

upper limit = 117.2, lower limit = 92.8 ✗

e The bars were accurately plotted on the graph (not shown), for 1 mark. The candidate has failed to use t in the calculation, so it is not known if a correct value would have been taken from the Statistics Sheets.

(b) The mean for C is the most reliable ✓ since the limits are the narrowest ✓.

e This is correct. The candidate scores both marks.

(c) There is a difference ✗ since the limits to B are so wide compared with those for C ✗.

e The candidate does not understand *'significant'* difference. The role of confidence limits in determining reliability has been confused with their role in indicating significant difference between sample means.

(d) (i) There is no difference between the mean mass of apples of the two varieties ✗.

e There is a difference. The null hypothesis is about there being no *significant* difference.

(ii) 58 ✓

e This is correct, for 1 mark.

(iii) $0.002 > p > 0.001$ ✓

e This is correct, for 1 mark.

(iv) The null hypothesis is rejected ✓.

e This is correct, for 1 mark.

(e) Variety D is larger than variety A since its cells are larger to accommodate the greater number of chromosomes ✓.

e This is correct, for 1 mark.

Candidate B

(a) 95% confidence limits = 105 ± (2.045 ✓ × 12.2)

upper limit = 130, lower limit = 80 ✓

e The bars were accurately plotted on the graph (not shown), for 1 mark.

(b) The mean for variety C has the narrowest limits ✓ which indicates that it is the most reliable ✓, i.e. closest to the true mean.

(c) The means do not appear to be significantly different ✓ since the confidence limits overlap ✓.

(d) (i) That any difference between the means of the samples is simply due to chance ✓.

(ii) d.f. = 58 ✓

(iii) $0.002 > p > 0.001$ ✓

(iiv) The difference is highly significant ($0.002 > p > 0.001$) ✓

(e) Variety D ($3n$) has apples which are approximately 50% bigger than variety A ($2n$) ✓.

This excellent answer from Candidate B scores full marks.

Overall, Candidate A scores 7 marks and Candidate B scores 12.

Question 7

The genetics of the budgerigar have been extensively studied, particularly with regard to colour variants.

Wild-type budgerigars are light green, determined by the allele B. The allele b is recessive to B and, in the homozygous condition, determines sky-blue (genotype bb).

The B/b gene interacts with the sex-linked *ino* gene (X^+/Xi). When the dominant wild-type allele (X^+) is present, the alleles B and b are expressed as above. However, the Xi allele can influence the action of the B/b gene. XiXi (or XiY) along with the B allele determines yellow, while in combination with the b allele produces an albino condition.

Sex is determined in budgerigars by the X and Y chromosomes, though females are XY and males XX.

(a) Using the symbols shown above, list all the genotypes of the following:
- **light green (wild-type) male**
- **yellow female**
- **albino male** (4 marks)

(b) Two budgerigars of the following genotypes were crossed.

BbX^+Y × bbX^+Xi

Determine the expected proportions of offspring produced in terms of colour and gender. Show your working in a genetic diagram. (5 marks)

(c) Explain how the genotype of a light green female budgerigar (BBX^+Y or BbX^+Y) may be determined. (2 marks)

Total: 11 marks

Candidates' answers to Question 7

Candidate A

(a) Light green male: BBX^+X^+ ✗

Yellow female: BBXiY ✗

Albino male: bbXiXi ✓

Light green males and yellow females have more than one possible genotype. The single genotype presented for each is not worth a mark. Always check that your answer is as complete as possible.

(b)

	bX^+	**bXi**
BX^+	BbX^+X^+ Light green male	BbX^+Xi Light green male
bY	bbX^+Y Sky blue female	bbXiY Albino female

Female gametes ✗
Male gametes ✓
Appropriate Punnett square ✓
Genotypes correct for gametes ✓
Phenotypes correct for genotypes ✓

The female gametes are incomplete — the law of independent assortment has not been applied — and so not worthy of a mark. However, this mistake is not penalised further.

(c) This would require a test cross with the double recessive ✓.

This is not complete since alternative results are not provided.

Candidate B

(a) BBX^+X^+, BbX^+X^+ ✓, BBX^+Xi, BbX^+Xi ✓

BBXiY, BbXiY ✓

bbXiXi ✓

(b)

	bX^+	**bXi**
BX^+	BbX^+X^+ Light green male	BbX^+Xi Light green male
BY	BbX^+Y Light green female	BbXiY Yellow female
bX^+	bbX^+X^+ Sky blue male	bbX^+Xi Sky blue male
bY	bbX^+Y Sky blue female	bbXiY Albino female

Female gametes ✓
Male gametes ✓
Appropriate Punnett square ✓
Genotypes correct ✓
Phenotypes correct ✓

(c) Test cross with bbX^+X^+ ✓. If some offspring are blue then the parent is Bb whereas if they are all green then the parent is BB ✓.

Candidate B scores full marks for this excellent answer.

Overall, Candidate A scores 6 marks and Candidate B scores 11.

Question 8

The two-spot ladybird (*Adalia bipunctata*) is typically red–orange with black spots but a melanic form exists which is black with four red spots. The ten-spot ladybird (*Adalia decempunctata*) is typically orange with black spots but a melanic form also exists. These are shown diagrammatically below.

Adalia bipunctata

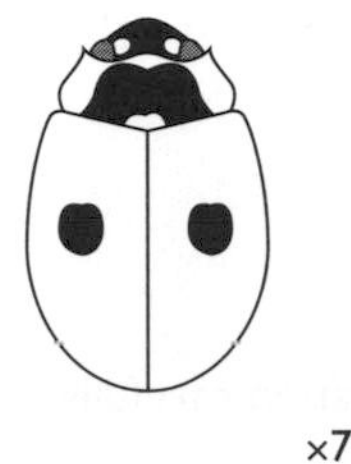

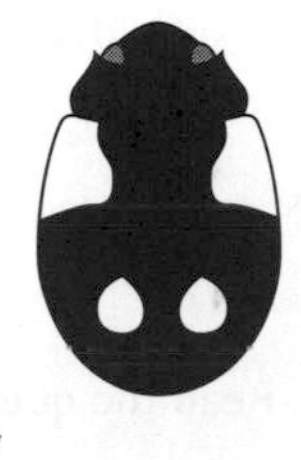

×7

Adalia decempunctata

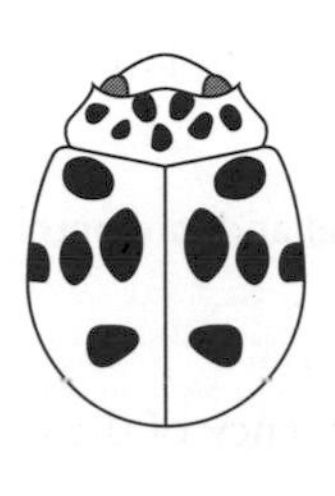

×7

(a) Explain why the two-spot and ten-spot ladybirds would be regarded as different species and why the melanic forms are simply variants within each species. (2 marks)

(b) Melanic forms are considered to be at an advantage under conditions of low levels of sunshine. The relative frequencies of melanic forms are certainly greater in more northerly regions. It is presumed that the black surfaces are more effective at absorbing radiant energy so that the body temperature rises rapidly and the ladybirds become more active.

With reference to natural selection, explain how populations of ladybirds in northerly regions possess relatively high frequencies of melanic forms. (4 marks)

(c) A population of the two-spot ladybird (*Adalia bipunctata*) was investigated in Carrickfergus. Out of a total of 500 sampled, 95 were found to be melanic. The melanic form is due to a dominant allele at a single gene locus.

(i) Calculate the proportion of the population that is represented by the typical (red–orange) form. (1 mark)

(ii) Calculate the relative frequency of the typical (red–orange) allele and the melanic (black) allele, assuming the population to be in Hardy–Weinberg equilibrium. (Show your working). (2 marks)

(iii) Calculate the relative frequencies of ladybirds that might be expected to be homozygous melanic and heterozygous melanic. (Show your working). (2 marks)

Total: 11 marks

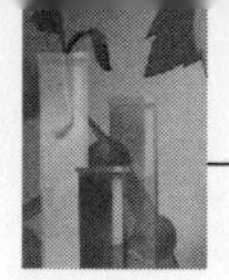

Candidates' answers to Question 8

Candidate A

(a) The two-spot and ten-spot ladybirds are different species because they cannot interbreed ✓. The melanic forms have the melanic genes ✗.

e The first point is correct, for 1 mark. The correct answer to why melanic forms are variants within each species is that variants of the same species are able to interbreed.

(b) In northerly regions the melanic form is more active and lays more eggs ✓ so is selected for.

e This lacks any detail and is worth only 1 mark.

(c) (i) $\frac{95}{500} = 19\%$ ✗

e No, this is the frequency of the melanic form. Read the question carefully!

(ii) $q^2 = 81\%$, so $q = 9\% = 0.09$ ✗

so $p = 1 - 0.09 = 0.91$ ✓

e The candidate has worked with percentages and obtained the wrong q value (the root of 81% is 0.9). However, understanding that $p + q = 1$ is shown, for 1 mark.

(iii) frequency of homozygotes = $p^2 = 0.91^2 = 0.83$ ✓

frequency of heterozygotes = $2pq = 2 \times 0.91 \times 0.09 = 0.16$ ✓

e Even though the values of p and q are wrong (already penalised), the candidate shows understanding of the frequencies of homozygotes and heterozygotes and obtains 2 marks.

Candidate B

(a) Two-spot ladybirds have red and black forms that are able to interbreed ✓. This is also the case for the 10-spot ladybird. However, forms of the two-spot ladybird are unable to interbreed with forms of the 10-spot ladybird ✓, so two-spot and ten-spot ladybirds are separate species.

(b) The black forms are better at absorbing solar radiation and so are more able to catch prey ✓, such as aphids. Thus females have more food for egg production, while males are more active in mating, i.e. they have greater fertility ✓, and so more of their genes are transmitted to the next generation ✓. The frequency of the melanic allele increases in the northerly populations ✓.

(c) (i) frequency of red form = $\frac{405}{500} = 81\% = 0.81$ ✓

(ii) $q^2 = 0.81$, so q (frequency of red allele) = 0.9 ✓

so p (frequency of black allele) = 0.1 ✓

(iii) frequency of homozygous melanic = $p^2 = 0.01$ ✓

frequency of heterozygous melanic = $2pq = 0.18$ ✓

Candidate B scores full marks.

Overall, Candidate A scores 5 marks and Candidate B scores 11.

Section A total for Candidate A: 41 marks out of 72

Section A total for Candidate B: 66 marks out of 72

Section B

Quality of written communication is awarded a maximum of 2 marks in this section. (2 marks)

Question 9

(a) Write an essay on how food is obtained and digested in each of the following animal phyla.

- **Cnidaria**
- **Platyhelminthes**
- **Annelida** (12 marks)

(b) Describe four trends in the evolution of the above animal phyla. (4 marks)

Total: 18 marks

Candidates' answers to Question 9

Candidate A

(a) Cnidaria: all animals in this phylum are aquatic, e.g. *Hydra*. They paralyse their live prey using stinging cells ✓. The captured food is then carried to the mouth by their tentacles ✓. These animals have a simple sac-like gut ✓. To start with digestion is extracellular but in the later stages it is intracellular ✓.

Platyhelminthes: all animals in this phylum are triploblastic and unsegmented, e.g. planaria. Food is ingested by the eversion of a muscular pharynx ✓. Animals in this phylum have a sac-like gut ✓. At the start digestion is extracellular but the animal subsequently digests intracellularly ✓.

Annelida: all animals in this phylum are soft-bodied and are either aquatic or damp terrestrial, e.g. earthworm. These animals have a canal-like gut, i.e. there is a mouth and an anus ✓. All digestion occurs extracellularly ✓.

(b) There is evolution from bilateral symmetry to radial symmetry ✗. There is a change from the diploblastic to the triploblastic condition ✓, with the mesoderm used

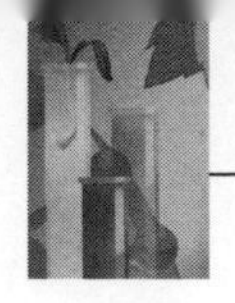

to produce more tissues and organs. Digestion changes so that it is completely extracellular ✓, i.e. no intracellular. There is a change from a sac-like gut towards a canal-like gut ✓.

The answer to part (a) has nine appropriate points and the answer to part (b) has three correct points. Note that the trend in symmetry is from radial to bilateral. The candidate expresses ideas clearly in a well-sequenced account and so gains 2 marks for QWC.

Candidate B

(a) Cnidaria, e.g. *Hydra*, possess stinging cells called cnidocysts that are used to paralyse live prey ✓. The impaled and paralysed prey is then transferred via the mouth into the enteron ✓. Here enzymes are released so that digestion takes place initially extracellularly ✓. However, partially digested food is taken up by endocytosis ✓ so that digestion is completed intracellularly ✓.

Platyhelminths, e.g. planarians, ingest food items by the eversion of a muscular pharynx ✓ which are then taken into a branched gut ✓. Digestion is initially extracellular ✓ though completed intracellularly ✓. Undigested material is egested through the mouth (✓).

Annelida, e.g. earthworms, feed on detritus. The gut is a canal with a mouth and an anus ✓. The one-way nature of the gut allows specialisation along the gut ✓. For example, there is a crop for the storage of food ✓. Digestion is completed extracellularly (✓).

(b) There are a number of evolutionary trends in the animal kingdom. Organisms become more complex ✗. They also become larger ✗. There are changes in the structure of the gut ✗ and also in way digestion occurs ✗.

The answer to part (a) is a full account of feeding in the three phyla and scores the full 12 marks. However, the answer to part (b) shows a relative weakness. The candidate fails to score here because the points made are too general and lack detail. However, on the basis of (a), this is a fluent and well-structured account, with good use of appropriate terminology and so 2 marks are awarded for QWC. A total of 14 marks are awarded for the essay.

Overall, both candidates score 14 marks for the essay.

Section B total for Candidate A: 14 marks out of 18

Section B total for Candidate B: 14 marks out of 18

Paper total for Candidate A: 55 marks out of 90

Paper total for Candidate B: 80 marks out of 90